SOIL AND WATER

Physical Principles and Processes

PHYSIOLOGICAL ECOLOGY

A Series of Monographs, Texts, and Treatises

EDITED BY

T. T. KOZLOWSKI

University of Wisconsin
Madison, Wisconsin

T. T. KOZLOWSKI. Growth and Development of Trees, Volume I — 1971; Volume II — *in preparation.*

DANIEL HILLEL. Soil and Water: Physical Principles and Processes, 1971

Soil and Water

Physical Principles and Processes

DANIEL HILLEL

DEPARTMENT OF SOIL SCIENCE
THE HEBREW UNIVERSITY OF JERUSALEM
REHOVOT, ISRAEL

ACADEMIC PRESS *New York and London*

ACADEMIC PRESS, INC.
111 Fifth Avenue, New York, New York 10003

United Kingdom Edition published by
ACADEMIC PRESS, INC. (LONDON) LTD.
Berkeley Square House, London W1X 6BA

LIBRARY OF CONGRESS CATALOG CARD NUMBER: 79-127685

PRINTED IN THE UNITED STATES OF AMERICA

*Dedicated to the memory of
my father, whose silent
presence has guided my effort.*

Contents

Part II: THE FIELD WATER CYCLE

6. Infiltration-Entry of Water into Soil

7. Redistribution of Soil Moisture following Infiltration

8. Groundwater Drainage

9.   Evaporation from Bare-Surface Soils

10. Uptake of Soil Water by Plants

11. Water Balance and Energy Balance in the Field

Appendix 1.   Numerical Solution of the Flow Equation

Contents

Preface

Soil and water are the two fundamental resources of our natural environment as well as of our agriculture. The increasing pressure of population has made these resources scarce or has led to their abuse in many parts of the world. The necessity to husband and manage these resources efficiently on a sustained basis is one of the most vital tasks of our age. For this reason, it has become increasingly important to deepen and disseminate knowledge of the properties and behavior of the soil–water system in relation to climatological conditions, plant growth, and the hydrological cycle.

In this book I attempt to describe the physical principles governing the soil–water system and particularly the sequence of processes constituting the cycle of water in the field. The presentation is meant for students and for professional workers in soil physics and other related disciplines (such as botany, ecology, agronomy, microbiology, geology, hydrology, geography, as well as agricultural and civil engineering) who need or might be interested in a fundamental and up-to-date exposition of soil physics.

Although the presentation is oriented toward the applied aspects of soil physics, the approach is more fundamental than directly utilitarian. The author believes that the development of a fundamental approach to the solution of problems is, in the long run, the most practical way to teach and apply such a subject. A basic understanding of physical principles and processes will enable workers in the field to adjust their thinking to changing situations and unforeseen problems. Ready-made solutions, on the other hand, are necessarily specific and inflexible and will therefore rarely apply as new problems arise in varying circumstances.

An attempt was made to keep the discussion as readable as possible and to assume little or no previous knowledge of the subject matter. On the other hand, to avoid oversimplification, it was necessary to assume a basic knowledge of physics and mathematics. Readers or students who find difficulty in digesting the material for lack of sufficient background, as well as those interested in additional information, are referred to more fundamental texts and to numerous original papers. However, no attempt was made to compile an exhaustive or complete set of references. Rather, references are given selectively to illustrate the highlights of the topic as seen by the author. This is in keeping with the intended character of the book, which is expository rather than encyclopedic.

This book is an outgrowth of my experience in teaching soil physics and soil–water relationships on various levels to engineers, agronomists, botanists, geologists, and hydrologists, both in Israel and in the United States. The final impetus to transmute lecture notes into book form arose during my sojourn as Visiting Professor of Soil Physics at the University of Wisconsin in 1968–69 and at the University of Georgia in early 1970. At that time, it became apparent that the need for a basic, readable, and reasonably comprehensive and up-to-date exposition of the soil–water system and its physical interactions is indeed universal. I do not pretend to have fulfilled this need completely. No particular book by one or even several authors is likely to suffice. The field of soil physics is too important, too complex and too active to be encapsulated in any one book, which necessarily represents a particular point of view.

It is my great pleasure to acknowledge the counsel and encouragement I received from colleagues at the University of Wisconsin during the preparation of this book: Dr. Lincoln E. Engelbert, formerly chairman of the Department of Soils, at whose behest the task was undertaken in the first place; Dr. Wilford R. Gardner, my mentor and brother-in-spirit, who read and criticized the original version; Dr. Edward E. Miller, whose generous investment of time and numerous suggestions were most helpful; Dr. Champ B. Tanner, whose personal example of dedication sustained me when my resolve might have faltered; Dr. Richard C. Amerman, of the A.R.S., U.S. Department of Agriculture, who also made numerous helpful comments, as did Dr. P. A. C. Raats of the A.R.S. and Dr. A. J. Peck of the C.S.I.R.O., Australia.

I am particularly indebted to Dr. T. T. Kozlowski, whose rigorous editing ameliorated what must have been a rather disheveled manuscript. Dr. A. Bertrand, head of the Agronomy Department at the University of Georgia, also read the manuscript and encouraged me to publish it. Other colleagues, too numerous to mention, helped me in many direct and indirect ways. They deserve to share the credit for whatever good features the book may contain, though I bear sole responsibility for all its shortcomings.

Madison, Wisconsin   DANIEL HILLEL

SOIL AND WATER

Physical Principles and Processes

Introduction

The importance of the soil–water system in nature and in the life of man has been realized since the dawn of civilization and man's awakening awareness of his relationship to his environment. The ancient Greeks considered soil and water to be two of the four primary elements composing all of nature. The ancient Hebrews maintained that man himself was created out of, and destined to return to "affar," which is, literally, the material of the soil. Today, no less than in ancient times, man is ultimately dependent for his subsistence upon the soil–water system and the plant life which it supports.

Despite the early realization, however, the actual science of soil physics, being the study of the state and transport of matter and energy in the soil, is a very young one. Historically, the development of other branches of soil science, namely, pedology (the study of the origin, development, and classification of soil as a body in nature) and soil chemistry (the study of the chemical composition and processes of the soil) preceded soil physics by a generation or two. Similarly, the early development of water science, or hydrology, was not accompanied by, nor based upon, a proper study of the soil-water system and the soil–water interactions which constitute so essential a link in the chain of processes comprising the hydrologic cycle.

Since the beginning of this century, however, and particularly since the 1930's, soil physics has become established and recognized increasingly as a vital field of universal interest, both as a separate discipline and as a meeting ground among the natural sciences. Today, indeed, research on various fundamental and applied aspects of soil physics is being carried out at hundreds of locations throughout the world, in connection with hydrology, ecology, engineering, and agriculture. Though much progress has been achieved and the science of soil physics is becoming increasingly exact and

1

quantitative, misunderstandings and misconceptions still abound. This is due primarily to the inherent difficulty of the subject.

The soil itself is of the utmost complexity. It consists of numerous solid components (mineral and organic) irregularly fragmented and variously associated and arranged in an intricate geometric pattern that is almost indefinably complicated. Some of the solid material consists of crystalline particles, while some consists of amorphous gels which may coat the crystals and modify their behavior. The adhering amorphous material may consist of iron and aluminum oxides, or of organic matter which attaches itself to soil particles and joins them together. The solid phase interacts with the fluids (water and solutes, gases) which permeate soil pores.

The soil–water complex does not exhibit constant properties or conditions of stable equilibrium, as it alternately wets and dries, saturates and de-saturates, swells and shrinks, disperses and flocculates, cracks, aggregates, compacts, and undergoes chemical changes and structural rearrangements. Particularly amenable to such changes are the colloidal clay particles, which adsorb water and exchange ions.

Water itself is a substance of unique and complex behavior, constantly dissolving or releasing materials, subject to frequent changes of state (solid, liquid, or vapor) and of properties (viscosity, surface tension, etc.), which are affected by temperature, pressure, and solutes.

Since the soil exhibits at all times exceedingly complex interactions among its constituents, it is well nigh impossible to define completely its exact physical state at any time. In dealing with any particular soil and condition, therefore, we are generally obliged to take the easy way out, which is to simplify our system by concentrating upon the factors which appear to have the greatest and most direct bearing upon the problem at hand, while dis-regarding as extraneous complications the factors which may seem to be of secondary importance.

In many cases, the theories and equations employed in soil physics do not describe the soil itself, but an ideal and well-defined model by which we simulate the soil. Thus, for example, at different times and for different purposes, the soil may be compared to a collection of small spheres, or to a bundle of capillary tubes, or to a collection of parallel colloidal platelets, or even to a mechanical continuum. We tend to describe the system macro-scopically rather than microscopically; that is, instead of defining the state of any particular particle or pore, we seek to characterize the whole by a gross averaging of its various parts.

The value of these models and representations depends upon the degree of their reality in any particular case, but even at best they cannot provide anything but a partial explanation of soil behavior. The complications which we may choose to disregard do not in fact disappear. Having once defined the most important (or "primary") effects, we find that to refine our model we must now consider the next-to-the-most-important ("secondary") effects, and so *ad infinitum*. Our developing knowledge of the soil, as of other complex systems, is achieved by successive approximations.

In applying to the soil oversimplified concepts and theories borrowed from simpler or "purer" systems, we must be careful not to take our models too seriously or literally. Present-day theories of soil physics should therefore be taken with a grain or two of salt, as they were developed by entirely fallible (though courageous) soil physicists desperately attempting to make their system manageable by simplifying it. As the science develops, however, its tools are becoming more sophisticated and capable of handling some of the complexities which previous soil physicists perforce disregarded.

The use of models, of course, is not unique to soil physics. It is, in fact, a traditional and indispensable tool of all physical science. A theory is a language model of a process or structure of the material world. The model attempts to explain how things behave, and the causal links between observed events.

To be sure, all models and theories are idealizations; they may not correspond to the observables in an obvious one-to-one way. No one, for example, has even seen an electron, yet the conceptual model has played an enormously important role in physics. Heat, for another example, does not really "flow," but the analogy to a fluid helps us to grasp the behavior of heat, to see it in familiar terms, and to facilitate our speculations. The visual image of something flowing leads us to think of gradients, of a natural tendency to flow from a higher to a lower level. This has turned out to be useful and to correspond with facts. However, when a model begins to mislead, to depart too grossly from the facts, we must modify or replace it.

Models are best expressed in the concise and terse language of mathematics. We set an equation which describes the model's behavior. We can transform the equation to anticipate how the model should behave under different conditions. In this way, the model serves not only to summarize what we know, but also to predict what we still do not know. Then, we check our prediction by experiment, and if it fits, we have a working model. Thus, theory cannot advance without experimentation. Conversely, experimentation without theory is likely to be sterile and pointless, as it might hopelessly bog us down in an ever-increasing mire of seemingly unrelated and random facts.

So much for the philosophy of our science. Our actual present-day knowledge, both empirical and theoretical, of the physical behavior of the soil–water system is still rather fragmented and uncoordinated. There is a large mass of data which still contains contradictions and still awaits the development of integrating theory. In describing the moot and unresolved questions, the author has no choice but to draw upon his own judgment. This is a rather risky task, since no involved individual is ever entirely without bias. For this reason, the reading of this book must be done critically. A valid book on soil physics must reflect the complexity of the system even while attempting to present a coordinated and logical description of our present-day in-complete knowledge of it.

For general background and supplementary study of various aspects of soil science, the reader is referred to the following books: "Chemistry of Soil" (Bear, 1955); "Clay Mineralogy" (Grim, 1953); "The Physical Chem-istry and Mineralogy of Soils" (Marshall, 1964); "Fundamentals of Soil Science" (Millar *et al.*, 1965); " Irrigation Principles and Practices" (Israelson and Hansen, 1962); " Soil Conservation" (Kohnke and Bertrand, 1959); " Soil Plant Relationships" (Black, 1968); "Soil Physics " (Baver, 1956); "The Nature and Properties of Soils " (Buckman and Brady, 1960); " Soil," U.S.D.A. Yearbook of Agriculture (1957); "Water," U.S.D.A. Yearbook of Agriculture (1955); "Soil Physical Conditions and Plant Growth" (Shaw, ed., 1952); "Methods of Soil Analysis" (Black, ed., 1965); "Irrigation of Agricultural Land" (Hagan *et al.*, eds., 1967); and "Agricultural Physics" (Rose, 1966).

Part I: *PHYSICAL PRINCIPLES*

1 *Basic Physical Properties of Soils*

A. General

The term "soil" refers to the weathered and fragmented outer layer of the earth's land surface. It is formed initially from disintegration and decomposition of rocks by physical and chemical processes, and is influenced by the activity and accumulated residues of numerous biological species. The soil can be studied and described from many different points of view, and hence soil science is in fact a conglomeration of several separate, though interdependent, disciplines.

Our treatment of the soil is from the point of view of soil physics, which we can describe as the branch of soil science dealing with the physical properties of the soil, as well as with the description, measurement, and control of the physical processes taking place in the soil. As physics deals with matter and energy (their forms and interrelations), so soil physics deals with the state and movement of matter and with the fluxes and transformations of energy in the soil.

On the one hand, the study of soil physics aims at achieving a basic understanding of the soil and its role in the overall geophysical system of the earth's surface, with all its interrelated and cyclic processes (such as the water cycle and energy exchange). On the other hand, the practice of soil physics aims at providing the tools for the proper management of the soil by means of irrigation, drainage, soil and water conservation, soil tillage, soil structure improvement, soil aeration, and the regulation of soil heat, as well as for the use of the soil as a building material or as a foundation for roads or structures.

Soil physics is thus seen to be both a basic and a practical science with a very wide range of interests, many of which are shared by other branches of soil science and by other varied sciences such as hydrology, climatology, agronomy, botany, ecology, and geology.

A soil which contains adequate amounts of the various substances required in plant nutrition can be considered to possess "chemical fertility." Such fertility, though essential, does not by itself ensure the success of crops. The suitability of a soil as a medium for plant growth depends not only upon the presence and quantity of chemical nutrients, but also upon the state and movement of water and air and upon the mechanical attributes of the soil and its thermal regime. The soil must be loose and sufficiently soft and friable to permit root development without mechanical obstruction. The pores of the soil should be of the volume and size distribution that will ensure adequate movement and retention of water and air to meet plant needs. In short, in addition to "chemical fertility," the soil should also possess "physical fertility," with both attributes equally essential for overall soil productivity.

B. Soil as a Disperse Three-Phase System

Natural systems can consist of one or more substances and of one or more phases. A system consisting of a single substance is also monophasic if in all its parts the properties are similar. An example of such a system is a body of water consisting entirely of uniform ice. Such a system is homogeneous.

A system of uniform chemical composition may also be heterogeneous if the substance exhibits different properties in various regions of the system. A region inside a system which is internally uniform in physical properties is called a phase. A mixture of ice and water, for instance, is chemically uniform, but physically heterogeneous, consisting of two phases.

The three ordinary phases in nature are the solid, liquid, and gaseous phases. A system including several substances can also be monophasic. For example, a solution of salt and water is a homogeneous liquid. A system of several substances can obviously also be polyphasic.

In a heterogeneous system, the properties differ not only between one phase and another, but also between the internal parts of each phase and the boundaries or interfaces of the phase with its neighboring phase or phases. Interfaces exhibit specific phenomena, such as adsorption, surface tension, and friction, which result from the interaction of adjacent phases and therefore do not exist within the homogeneous phases themselves. The importance of these phenomena in the overall system is proportional to the size of the interfacial area per unit volume of the system.

Disperse systems are those in which at least one of the phases is subdivided into minute particles which together exhibit a very large surface area. Examples of frequently encountered disperse systems are the colloidal sols, gels, emulsions, aerosols, etc.

The soil is a heterogeneous, polyphasic, particulate, disperse, and porous system, in which the interfacial area per unit volume can be enormously large. The disperse nature of the soil and its consequent interfacial activity give rise to such phenomena as swelling, shrinkage, dispersion, aggregation, adhesion, adsorption, ion exchange, etc.

The three phases of ordinary nature exist in the soil as well: the solid phase, consisting of soil particles; the liquid phase, consisting of soil water, which always contains dissolved substances, so that it should be called the soil solution; and the gaseous phase, consisting of soil air.

The soil is thus seen to be an exceedingly complex system. Its solid matrix consists of particles differing in chemical and mineralogical composition as well as in size, shape, and orientation. The mutual arrangement or organization of these particles in the soil determines the characteristics of the pore spaces in which water and air are transmitted or retained. The water and air also vary in composition, both in time and in space.

It is not always easy to separate these phases, as they interact very strongly upon one another. For the sake of expressing their quantitative proportions, however, we shall arbitrarily and schematically consider them as independent constituents.

C. Volume and Mass Relationships of Soil Constituents

Figure 1.1 gives a schematic diagram of a soil which will help us to define the volume and mass relationships of the three soil phases.

The complete histogram represents the total mass and volume of the soil and it is divided into three sections which are in general quantitatively unequal; the lower section represents the solid phase, the middle section the liquid phase, and the top section the gaseous phase.

The masses of these components are marked on the right-hand side: the mass of air M_a, which is negligible and usually considered to be zero; the mass of water M_w; the mass of solids M_s; and the total mass M_t. These masses are often substituted by the weights (being the product of the corresponding masses and the acceleration of gravity). The volumes of the same components are indicated on the left-hand side of the diagram: the volume of air V_a; the volume of water V_w; the volume of pores $V_f = V_a + V_w$; the volume of solids V_s; and the total volume V_t of the soil.

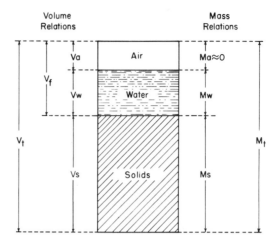

Fig. 1.1. Schematic diagram of the soil as a three-phase system.

On the basis of this diagram, we can now define terms which are generally used to express the quantitative interrelations of the three primary soil constituents:

1. DENSITY OF SOLIDS (MEAN PARTICLE DENSITY) ρ_s:

$$\rho_s = \frac{M_s}{V_s} \tag{1.1}$$

In most mineral soils, the mean density of the particles is about 2.6–2.7 gm/cm^3. The presence of organic matter lowers the value of ρ_s. Sometimes the density is expressed in terms of the *specific gravity*, being the ratio of the density of the material to that of water at 4°C and at atmospheric pressure. In the metric system, since the density of water is assigned the value of unity, the specific gravity is numerically (though not dimensionally) equal to the density.

2. DRY BULK DENSITY ρ_b:

$$\rho_b = \frac{M_s}{V_t} = \frac{M_s}{V_s + V_a + V_w} \tag{1.2}$$

The dry bulk density expresses the ratio of the mass of dried particles to the total volume of the soil (including the particles as well as the pores). It is obviously smaller than the value of ρ_s. For a soil in which the pores constitute half the volume, ρ_b is half of ρ_s, namely 1.3–1.35 gm/cm^3. In sandy soils, ρ_b can be as high as 1.6, whereas in aggregated loams and in clay soils, it can be as low as 1.1 gm/cm^3. The bulk density is affected by the structure of the soil, i.e., its looseness or degree of compaction, as well as its swelling and shrinkage characteristics, which are dependent upon the wetness. In extremely compacted and well-graded soils, the bulk density might approach, but never reach, the particle density. No matter how compacted, the particles cannot interlock perfectly and the soil remains a porous body, never completely impervious. In swelling soils, bulk density varies with moisture content (wetness).

3. TOTAL (WET) BULK DENSITY ρ_t:

$$\rho_t = \frac{M_t}{V_t} = \frac{M_s + M_w}{V_s + V_a + V_w} \tag{1.3}$$

This is an expression of the total mass of a moist soil per unit volume. The wet bulk density depends even more strongly than the dry bulk density upon the wetness or moisture content of the soil.

4. DRY SPECIFIC VOLUME v_b:

$$v_b = \frac{V_t}{M_s} = \frac{1}{\rho_b} \tag{1.4}$$

The volume of a unit mass of dry soil (cubic centimeter per gram) serves as another index of the degree of looseness or compaction of the soil.

5. POROSITY f:

$$f = \frac{V_f}{V_t} = \frac{V_a + V_w}{V_s + V_a + V_w} \tag{1.5}$$

The porosity is an index of the relative pore volume in the soil. Its value generally lies in the range 0.3–0.6 (30–60%). Coarse-textured soils tend to be less porous than fine-textured soils, though the mean size of individual pores is greater in the former than in the latter. In clayey soils, the porosity is highly variable as the soil alternately swells, shrinks, aggregates, disperses, compacts, and cracks. As generally defined, the term porosity refers to the volume fraction of pores, but this value should be equal, on the average, to the areal porosity (the fraction of pores in a representative cross-sectional area) as well as to the average lineal porosity (being the fractional length of pores intersected by a line passing through the soil in any direction). The total porosity, in any case, reveals nothing about the *pore size distribution*, which is itself an important property to be discussed in a later section.

6. VOID RATIO e:

$$e = \frac{V_a + V_w}{V_s} = \frac{V_f}{V_t - V_f} \tag{1.6}$$

The void ratio is also an index of the relative volume of soil pores, but it relates to the volume of solids rather than to the total volume of soil. The advantage of this index over the previous one is that a change in pore volume changes the numerator alone, whereas a change of pore volume in terms of the porosity will change both the numerator and denominator of the defining equation. Void ratio is the generally preferred index in soil engineering and mechanics, whereas porosity is the more frequently used index in agricultural soil physics. Generally, e varies between 0.3 and 2.0.

7. SOIL WETNESS

The wetness, or relative water content, of the soil can be expressed in various ways: relative to the mass of solids, relative to the total mass, relative to the volume of solids, relative to the total volume, and relative to the volume of pores. The most commonly used indexes are defined in the following paragraphs.

7a. Mass wetness w:

$$w = \frac{M_w}{M_s} \tag{1.7}$$

This is the mass of water relative to the mass of dry soil particles, often referred to as the "gravimetric water content." The term "dry soil" is generally defined as a soil dried to equilibrium in an oven at 105°C, though clay will often contain appreciable quantities of water at that state of dryness and even at higher temperatures. Soil dried in "ordinary" air will generally contain several per cent more water than oven-dry soil, a phenomenon due to vapor adsorption and often referred to as soil "hygroscopicity." In a mineral soil that is saturated, w can range between 25% and 60% depending on the bulk density. The saturation water content is generally higher in clayey than in sandy soils. In organic soils, such as peat or muck, the saturation water content on the mass basis can sometimes exceed 100%.

7b. Volume wetness θ:

$$\theta = \frac{V_w}{V_t} = \frac{V_w}{V_s + V_f} \tag{1.8}$$

The volume wetness (often termed "volumetric water content") is generally computed on the basis of the total volume of the soil rather than on the basis of the volume of particles alone. In sandy soils, the value of θ at saturation is on the order of 40–50%; in medium-textured soils, it is approximately 50%; and in clayey soils, it can be on the order of 60%. In the latter, the relative volume of water at saturation can exceed the porosity of the dry soil, since clayey soils swell upon wetting. The use of θ rather than of w to express water content is often more convenient because it is more directly adaptable to the computation of fluxes and water quantities added to soil by irrigation or rain and to quantities subtracted from the soil by evapotranspiration or drainage.

7c. *Degree of saturation θ_s:*

$$\theta_s = \frac{V_w}{V_f} = \frac{V_w}{V_a + V_w} \tag{1.9}$$

This index (sometimes called simply "saturation") expresses the volume of water present in the soil relative to the volume of pores. The index θ_s ranges from zero in dry soil to 100% in a completely saturated soil. However, 100% saturation is seldom attained, since some air is nearly always present and may become trapped in a very wet soil. This is not a good index for swelling soils, in which porosity changes with wetness.

8. THE AIR-FILLED POROSITY (RELATIVE AIR CONTENT) f_a:

$$f_a = \frac{V_a}{V_t} = \frac{V_a}{V_s + V_a + V_w} \tag{1.10}$$

This is a measure of the relative air content of the soil, and as such is an important criterion of soil aeration. The index f_a is related negatively to the degree of saturation θ_s ($f_a = f - \theta_s$).

9. ADDITIONAL INTERRELATIONS

From the basic definitions given, it is possible to derive the relation of the various expressions to one another. The following are several of the most useful interrelations.[1]

(1) Relation of porosity and void ratio:

$$e = \frac{f}{1-f}, \qquad f = \frac{e}{1+e} \tag{1.11}$$

[1] The actual derivation of these and other interrelations is a useful exercise for students.

(2) Relation of degree of saturation to volume wetness:

$$\theta_s = \frac{\theta}{f}$$ (1.12)

(3) Relation of porosity to bulk density:

$$f = \frac{\rho_s - \rho_b}{\rho_s} = 1 - \frac{\rho_b}{\rho_s}$$ (1.13)

(4) Relation of mass to volume wetness:

$$\theta = \frac{w\rho_b}{\rho_w}$$ (1.14)

where ρ_w is the density of water (M_w/V_w). Since the bulk density ρ_b is generally greater than ρ_w, it is obvious that the volume wetness is greater than the mass wetness.

(5) Relation of air to water content:

$$f_a = f - \theta = f(1 - \theta_s)$$ (1.15)

Of the various parameters defined, the ones most commonly used in characterizing the physical relationships of soil constituents are the porosity f, bulk density ρ_b, and volume wetness θ.

D. Soil Texture

The primary particles in the soil may differ widely in size. Some are coarse enough to be seen easily with the naked eye, while others are small enough to exhibit colloidal properties. The term "soil texture" is an expression of the predominant size, or size range, of the particles, and it has both qualitative and quantitative connotations. Qualitatively, it refers to the "feel" of the soil material, whether coarse and gritty, or fine and smooth. An experienced soil classifier can tell by kneading or rubbing soil with his fingers whether it is coarse-textured or fine-textured.[2] Quantitatively, soil texture refers to the relative proportions of various sizes of particles in a given soil. The traditional method of characterizing particle sizes in soils is to divide these particles into

---

[2] The expressions "light soil" and "heavy soil" are used in common parlance to characterize the general physical behavior of different soils. Since a coarse-grained, sandy soil tends to be loose, well aerated, and easy to cultivate, it is called "light." A fine-textured soil, on the other hand, tends to absorb much water and become plastic and sticky when wet, as well as tight, compact, and cohesive when dry. Thus, it is called a "heavy" soil. These are unfortunate expressions, however, since in actual fact coarse-textured soils are generally more dense, i.e., have a lower porosity, than fine-textured soils, and thus are heavier, rather than lighter, in weight.

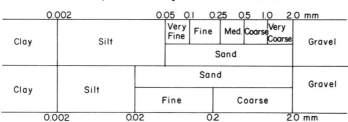

Fig. 1.2. Textural classification of soil fractions according to particle diameter ranges (given in logarithmic scale).

three size ranges known as *textural fractions,* or *separates: sand, silt,* and *clay.* There is as yet no universally accepted definition of these fractions. Two of the most prevalent classification schemes are shown in Fig. 1.2.

The overall textural designation, or *class,* is determined on the basis of mass ratios of these three fractions. Soils with different proportions of sand, silt, and clay are given different designations, as shown in the triangular diagram of Fig. 1.3.

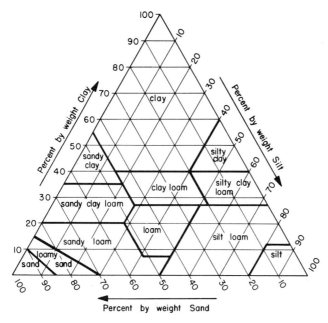

Fig. 1.3. Textural triangle, showing the percentages of clay (below 0.002 mm), silt (0.002–0.05 mm), and sand (0.05–2.0 mm) in the basic soil textural classes.

This method of classification is rather arbitrary. A possibly better method of characterizing soil texture is to show the continuous distribution of particle sizes, as in Fig. 1.4.

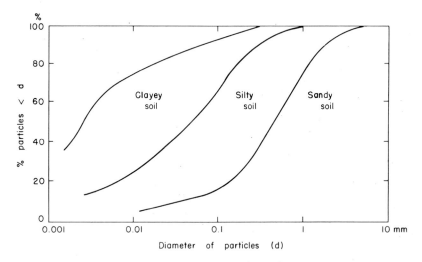

Fig. 1.4. Particle-size distribution in three types of soil (schematic).

Some soils are "well graded," i.e., have a continuous array of particles of various sizes. Other soils are "poorly graded," as they contain a preponderance of particles of one or several distinct size ranges.

E. Mechanical Analysis

Determination of particle size distribution, also known as the mechanical composition of the soil, is called *mechanical analysis*. Separation of particles into size groups can generally be carried out by sieving through graded sieves, down to a particle diameter of approximately 0.05 mm. To separate and classify still finer particles, the method of sedimentation is generally used. This consists of dispersing a sample of the soil in an aqueous suspension, and of measuring the settling velocity of the particles (or the density of the suspension from which the particles are settling).

According to Stokes's law,[3] the terminal velocity of a spherical particle

[3] A particle falling in a vacuum will encounter no resistance as it is accelerated by gravity and thus its velocity will increase as it falls. A particle falling in a fluid, on the other hand, will encounter a frictional resistance proportional to the product of its radius r, and velocity, and to the viscosity of the fluid.

The resisting force due to friction F_r was shown by Stokes (1851) to be

$$F_r = 6\pi v r u$$

settling under the influence of gravity in a fluid of a given density and viscosity is proportional to the square of the radius,

$$u = \frac{2}{9} \frac{r^2 g}{v} (\rho_s - \rho_f) \qquad (1.16)$$

where u is the settling velocity, r the particle radius, v the fluid viscosity, g the gravitational acceleration, and ρ_s and ρ_f the densities of the solid particle and of the fluid, respectively. The use of Stokes's law for measurement of soil-particle sizes is dependent upon certain simplifying assumptions which may not accord with reality (e.g., that the particles are spherical, of uniform density, and settle independently of each other; that the flow of the fluid around them is laminar; that the particles are sufficiently large to be un-affected by the thermal motion of the fluid molecules).

In fact, we know that soil particles are not spherical, and some may be platelike. Hence, the diameter calculated from the settlement velocity does not necessarily correspond to the actual dimensions of the particle. Thus, the results of a mechanical analysis based on sieving may differ from those of a sedimentation analysis. Furthermore, soil particles are not all of the same density. Most silicates have ρ_s values of 2.6–2.7 gm/cm^3. However, certain iron oxides and other heavy minerals may have density values of 5 gm/cm^3 or more. For Stokes's law to be applicable, the primary soil particles, often naturally aggregated, must be dispersed and made discrete by removal of cementing agents (such as organic matter, iron oxides, colloidal silica, and calcium carbonate) and by deflocculating the clay. For all these reasons, the mechanical analysis of soils can be a rather tedious and com-plicated process (Day, 1965).

---

where v is the viscosity of the fluid and r and u are the radius and velocity of the particle. Initially, as the particle begins its fall, its velocity increases. Eventually, a point is reached at which the increasing resistance force equals the constant downward force, and the particle then continues to fall without acceleration, at a constant velocity known as the terminal velocity, u_t.

The downward force due to gravity F_g is

$$F_g = \tfrac{4}{3}\pi r^3 (\rho_s - \rho_f) g$$

where $\tfrac{4}{3}\pi r^3$ is the volume of the spherical particle, ρ_s is its density, ρ_f is the density of the fluid, and g is the acceleration of gravity.

Setting the two forces equal, we obtain Stokes's law:

$$u_t = \frac{2}{9} \frac{r^2 g}{v} (\rho_s - \rho_f) = \frac{d^2 g}{18v} (\rho_s - \rho_f)$$

where d is the diameter of the particle. Assuming that the terminal velocity is attained almost instantly, we can obtain the time t needed for the particle to fall through a height h:

$$t = \frac{18 h v :}{d^2 g (\rho_s - \rho_f)}$$

F. Behavior of Clay

The fraction which determines the physical behavior of the soil most decisively is the colloidal clay, since it exhibits the greatest specific surface area and is therefore most active in physicochemical processes. Clay particles adsorb water and thus cause the soil to swell and shrink upon wetting and drying (Grim, 1958). Most of them are negatively charged and form an electrostatic double layer with exchangeable cations. Sand and silt have relatively small specific surface areas and consequently exhibit comparatively little physicochemical activity. These fractions can be termed the "soil skeleton," while the clay, by a similar analogy, can be thought of as the "flesh" of the soil. Together, they form the *solid matrix* of the soil.

The term clay designates not merely a range of particle sizes, but a large group of minerals, some of which are amorphous, but many of which occur in the form of highly structured microcrystals of colloidal size. The clay fraction thus differs mineralogically, as well as in particle sizes, from sand and silt, which are composed mainly of quartz and other primary mineral particles which have not been transformed chemically into secondary minerals as is the case with clay (Jenny, 1935; Jackson, et al., 1948). The various clay minerals differ greatly in properties and prevalence. And while the measurement of soil texture, discussed in the previous two sections, does give an idea of the quantity of clay in the soil, it reveals very little of the specific character and activity of the clay.

The most prevalent clay minerals are the layered aluminosilicates. Their crystals are composed of two basic structural units (Grim, 1963; Marshall, 1964; Low, 1968), namely: a tetrahedron of oxygen atoms surrounding a central cation, usually Si^{4+}, and an octahedron of oxygen atoms or hydroxyl groups surrounding a larger cation, usually Al^{3+} or Mg^{2+}. The tetrahedra are joined at their basal corners and the octahedra are joined along their edges by means of shared oxygen atoms. Thus, tetrahedral and octahedral layers are formed (Fig. 1.5).

The layered aluminosilicate clay minerals are of two main types, depending upon the ratios of tetrahedral to octahedral layers, whether 1:1 or 2:1. In the 1:1 minerals like kaolinite, an octahedral layer is attached by the sharing of oxygens to a single tetrahedral layer. In the 2:1 minerals like montmorillonite, it is attached in the same way to two tetrahedral layers, one on each side. A clay particle is composed of multiple-stacked composite layers (or unit cells) of this sort, called lamellae.

The structure described is an idealized one. Typically, some substitutions, or *isomorphous replacements*, of Al^{3+} for Si^{4+} occur in tetrahedral layers, and substitutions of Mg^{2+} for Al^{3+} occur in the octahedral layers. Hence, internally unbalanced negative charges occur at different sites in the lamellae.

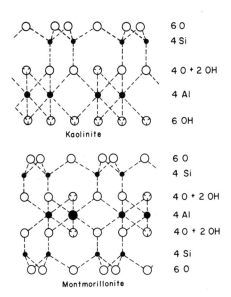

6 O
4 Si

4 O + 2 OH

4 Al

6 OH

Kaolinite

6 O
4 Si

4 O + 2 OH

4 Al

4 O + 2 OH

4 Si

6 O

Montmorillonite

Fig. 1.5. Schematic representation of the structure of aluminosilicate minerals.

Another source of unbalanced charge on clay minerals is the incomplete charge neutralization of terminal atoms on lattice edges. These charges are balanced externally by exchangeable ions (mostly cations), which concentrate near the external surfaces of the particle and occasionally penetrate into interlamellar spaces. These cations are not an integral part of the lattice structure, and can be replaced, or exchanged, by other cations. The cation-exchange phenomenon is of great importance in soil physics as well as soil chemistry, since it affects the retention and release of nutrients and salts, and the flocculation–dispersion processes of soil colloids.

A hydrated clay particle forms a colloidal micelle, in which the excess negative charge of the particle is neutralized by a spatially separated swarm of cations. Together, the particle surface and the neutralizing cations form an *electrostatic double layer*. The cation swarm consists partly of a layer more or less fixed in the proximity of the particle surface (known as the Stern layer), and partly of a diffuse distribution extending some distance away from the particle surface. This distribution is illustrated schematically in Fig. 1.6. It results from an equilibrium between two opposing effects: the Coulomb (electrostatic) attraction of the clay particle, versus the Brownian (kinetic) motion of the liquid molecules, inducing outward diffusion of the cations toward the intermicellar solution. Just as cations are adsorbed positively toward the clay particles, so anions are repelled, or adsorbed negatively, and relegated from the micellar to the intermicellar solution (Kruyt, 1949).

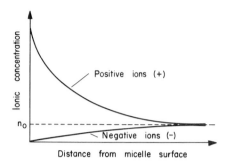

Fig. 1.6. Distribution of positive and negative ions in solution with distance from the surface of a clay micelle bearing net negative charge. Here, n_0 is the ionic concentration in the bulk solution outside the electrical double layer.

The quantity of cations adsorbed on soil-particle surfaces per unit mass of the soil under chemically neutral conditions is nearly constant and independent of the species of cation, and is generally known as the *cation exchange capacity*. Soils vary in cation exchange capacity from nil to perhaps 0.60 mEq per gm (Bear, 1955).

Clay minerals differ somewhat in *surface charge density* (i.e., the number of exchange sites per unit area of particle surface), and differ greatly in *specific surface area*. Hence, they differ also in their total *cation-exchange capacity*. Montmorillonite, with a specific surface area of nearly 800 m²/gm, has a cation-exchange capacity of about 0.95 mEq/gm, whereas kaolinite has an exchange capacity of only about 0.04–0.09 mEq/gm. The greater specific surface area of montmorillonite is due to its lattice expansion and consequent exposure of internal (interlamellar) surfaces, which are not so exposed in the case of kaolinite. Other clay minerals (such as illite, micas, palygorskite, etc.) often exhibit properties intermediate between those of kaolinite and montmorillonite.

The attraction of a cation to a negatively charged clay micelle generally increases with increasing valency of the cation. Thus, monovalent cations are replaced more easily than divalent or trivalent cations. Highly hydrated cations, which tend to be farther from the surface, are also more easily replaced than less hydrated ones. The order of preference of cations in exchange reactions is generally as follows (Jenny, 1932, 1938):

$$Al^{3+} > Ca^{2+} > Mg^{2+} > K^+ > Na^+ > Li^+$$

When confined clays are allowed to sorb water, swelling pressures develop, which are related to the osmotic pressure difference between the double layer and the external solution (Aylmore and Quirk, 1959). Depending upon their state of hydration and the composition of their exchangeable cations, clay particles may either flocculate or disperse (Jenny and Reitemeier, 1935). Dispersion generally occurs with monovalent and highly hydrated cations (e.g., sodium). Conversely, flocculation occurs at high solute concentrations and/or in the presence of divalent and trivalent cations (e.g., Ca^{2+}, Al^{3+}) when the double layer is compressed so that its repulsive effect is lessened and any two micelles can approach each other more closely.) Thus, the short-range attractive forces (known as the London–van der Waals forces) can come into play and join the individual micelles into *flocs*.

When a dispersed clay is dehydrated, it forms a dense and hard mass, or crust. On the other hand, when flocculated clay is dehydrated, it forms a crumbly and loose assemblage of small aggregates. Under rainfall action in the field, the dispersed clay will tend to become muddy, less pervious, and more highly erodible than flocculated clay. Thus, the desirable condition of a clayey soil is the flocculated one. Flocculation alone does not create an optimal structure, however, as will be explained in Section H.

G. Specific Surface and Adsorption Phenomena

The specific surface of a soil can be defined as the total surface area of the particles per unit mass (a_m), or per unit volume of particles (a_v), or per unit volume of dry soil (a_b):

$$a_m = \frac{A_s}{M_s} \tag{1.17}$$

$$a_v = \frac{A_s}{V_s} \tag{1.18}$$

$$a_b = \frac{A_s}{V_t} \tag{1.19}$$

where A_s is the total surface area of a mass of particles M_s having a volume V_s and contained in a bulk volume V_t of soil.

The specific surface is commonly expressed in terms of square meters per gram, or per cubic centimeter. It depends in the first place upon the sizes of the soil particles. In sand, the specific surface may be less than 1 m^2/gm, whereas in a clay, as we pointed out in the preceding section, the specific surface may be as high as several tens or even several hundred square meters per gram.

The specific surface area depends also upon the shape of the particles. Flattened or elongated particles obviously possess a greater specific surface per mass than spherical or cubical particles of the same average mass. Since clay particles are generally platy, they contribute even more to the overall specific surface area of the soil than is indicated by their small size alone. In addition to their external surfaces, certain clay crystals exhibit internal surface areas, such as those which form when the open lattice of montmorillonite expands on imbibing water.

It is thus apparent that the total specific surface of a soil, consisting of both external and internal surfaces, depends on the type of clay as well as on its total amount. Since many of the attributes of the soil relate to interfacial surface phenomena, the specific surface of soil is a highly pertinent property to study, and its measurement (Mortland and Kemper, 1965) might help to provide a basis for evaluating and predicting soil behavior. The specific surface often correlates with such soil properties as cation-exchange capacity, availability of certain nutrients, swelling, retention of water at high suctions, and certain mechanical properties such as plasticity and strength. For this reason, it is probable that the measurement of soil specific surface, though not yet as common as the measurement of soil texture by the traditional methods, may eventually prove to be a more meaningful and pertinent index for characterizing a soil than are the percentages of sand, silt, and clay.

It is interesting to consider the specific surface areas of idealized particles having regular geometric shapes.

For a sphere of diameter d, the ratio of surface to volume is

$$a_v = \frac{\pi d^2}{\pi d^3/6} = \frac{6}{d} \tag{1.20}$$

and the ratio of surface to mass is

$$a_m = \frac{6}{\rho_s d} \tag{1.21}$$

Where the particles have a density ρ_s of about 2.65 gm/cm^3, we obtain, approximately,

$$a_m \approx \frac{2.3}{d} \tag{1.22}$$

For a cube of edge L, the ratio of surface to volume is

$$a_v = \frac{6L^2}{L^3} = \frac{6}{L} \tag{1.23}$$

and the ratio of surface to mass is, again,

$$a_m = \frac{6}{\rho_s L} \tag{1.24}$$

Thus, the expressions for particles of nearly equal dimensions, such as most sand and silt grains, are similar, and knowledge of the particle size distribution can allow us to calculate the approximate specific surface by the summation equation:

$$a_m = \frac{6}{\rho_s} \frac{\sum d_i^2}{\sum d_i^3} \tag{1.25}$$

where $6/\rho_s \approx 2.3$.

Now let us consider a platy particle. For the sake of argument, we can assume that our plate is square-shaped, with sides L and thickness l. The surface-to-volume ratio is

$$a_v = \frac{2L^2 + 4Ll}{L^2 l} \tag{1.26}$$

and the surface-to-mass ratio

$$a_m = \frac{2(L + 2l)}{\rho_s Ll} \tag{1.27}$$

If the platelet is very thin, so that its thickness l is negligible compared to principal dimension L, and if $\rho_s = 2.65$ gm/cm^3, then

$$a_m \approx \frac{2}{\rho_s l} \approx \frac{0.75}{l} \quad \text{cm}^2/\text{gm} \tag{1.28}$$

Thus, the specific surface area of a clay can be estimated if the thickness of its platelets is known. For example, the thickness of a platelet of fully dispersed montmorillonite is approximately 10 Å, or 10^{-7} cm. Therefore, $A_m \approx 0.75/10^{-7}$, or 750 m^2/gm, which compares closely with the measured value.

The standard method for measuring the specific surface of materials is by adsorption of a gas such as nitrogen. Easier methods, which give relative results, are based on retention of a polar organic molecule such as ethylene glycol or glycerol (Dyal and Hendricks, 1950; Bower and Goertzen, 1959; Sor and Kemper, 1959).

The adsorption phenomenon was described by de Boer (1953). At low gas pressures, the amount of a gas adsorbed per unit area of adsorbing surface, σ_a, is related to the gas pressure P, the temperature T, and the heat of adsorption Q_a by the equation

$$\sigma_a = k_i P \exp(Q_a/RT) \tag{1.29}$$

where R is the gas constant and k_i is also a constant. Thus, the amount of adsorption increases with pressure, but decreases with temperature.

The equation of Langmuir (1918) indicates the relation between the gas pressure P and the volume of gas adsorbed per gram of adsorbent, v, at constant temperature:

$$\frac{P}{v} = \frac{1}{k_2 v_m} + \frac{P}{v_m} \qquad (1.30)$$

where v_m is the volume of adsorbed gas which forms a complete monomolecular layer over the adsorbent, and can be obtained by plotting P/v vs. P. The specific surface of the adsorbent can then be calculated by determining the number of molecules in v_m and multiplying this by the cross-sectional area of these molecules. The Langmuir equation is based on the assumption that only one layer of molecules can be adsorbed, and that the heat of adsorption is uniform during the process.

Brunauer *et al.* (1938) derived what has come to be known as the BET equation, based on multilayer adsorption theory:

$$\frac{P}{v(P_0 - P)} = \frac{1}{v_m C} + \frac{(C-1)P}{v_m C P_0} \qquad (1.31)$$

where v is the volume of gas adsorbed at pressure P, v_m is the volume of a single layer of adsorbed molecules over the entire surface of the adsorbent, P_0 is the gas pressure required for monolayer saturation at the temperature of the experiment, and C is a constant for the particular gas, adsorbent, and temperature. The volume v_m can be obtained from the BET theory by plotting $P/v(P_0 - P)$ vs. P/P_0. The density of the adsorbed gas is usually assumed to be that of the liquefied or the solidified gas.

Polar adsorbents (such as water) may not obey the BET or Langmuir equations (which are similar at low pressures), since their molecules or ions may tend to cluster at charged sites rather than spread out evenly over the adsorbent surface. The use of various adsorbents and techniques for the measurement of the specific surface area of soil materials was described by Mortland and Kemper (1965).

H. Soil Structure

Soil structure is generally defined as the mutual arrangement, orientation, and organization of the particles in the soil. The term is also used sometimes with reference to the geometry of the pore spaces. Since the arrangement of soil particles is generally too complex to permit any simple geometric characterization, there is no practical way to measure soil structure directly. Therefore, the concept of soil structure is used in a qualitative sense. The methods which have been proposed for characterization of soil structure are in fact indirect methods which measure some attributes affected by the structure rather than the structure itself. Many of these methods are rather arbitrary.

Unlike soil texture and specific surface, which are more or less constant for a given soil, the structure is highly dynamic and may change greatly from time to time in response to changes in natural conditions, biological activity, and soil-management practices. Soil structure can be of decisive importance in determining soil productivity since it greatly affects the water, air, and heat regimens in the field. Soil structure also influences the mechanical properties of the soil, which may in turn affect seed germination, seedling establishment, and root growth. Moreover, soil structure can affect the performance of agricultural operations such as tillage, irrigation, drainage, and planting (Russell, 1938; Boekel, 1963).

In general, it is possible to recognize three types of soil structure—*single-grained*, *massive*, and *aggregated*. When the particles are completely unattached to each other, the structure is called "single-grained." When the particles are bonded in large and massive blocks, the structure can be called "massive." As against these two extremes, there is an intermediate condition in which the soil particles are organized in small clods known as "aggregates." Inside of these aggregates, the particles are attached more or less stably, by intra-aggregate bonds.

Aggregated structure can be characterized either qualitatively (Soil Survey Manual, 1951), by describing the typical shapes of the aggregates (e.g., cubic, columnar, platy) or quantitatively, by measuring their sizes (Kemper and Chepil, 1965). Such measurements can be made by either "dry-sieving" or "wet-sieving." The latter type of measurement, usually carried out while the aggregated sample is immersed in water, is used as an index of the stability of the aggregates toward the destructive or slaking action of water (Kemper, 1965).

Additional methods of characterizing soil structure are based on the size distribution of the pores, the mechanical properties of the soil, or the permeability of the soil to air and water (Reeve, 1965). None of these methods has been accepted universally. In each case, the choice of the method to be used depends upon the problem, the soil, the equipment available, and, not the least, upon the soil physicist. The results obtained from some of these methods (e.g., wet sieving: Yoder, 1936; Russell and Feng, 1947) depend on the techniques employed.

The formation and stability of soil aggregates is dependent largely upon the quantity and state of clay, and upon the presence of various forms of organic matter. Emerson (1959) described a model of soil crumbs based upon the various ways in which assemblages of clay particles ("clay domains" or flocs) associate and attach to quartz particles of sand and silt to form microaggregates and macroaggregates. The clay not only cements aggregates internally, but often also coats over natural aggregates (often called *peds*) to form *clay skins*.

Flocculation of soil clay is a necessary condition for aggregate formation. By itself, however, the process of clay flocculation does not create the macro-aggregates and macropores which are so necessary for adequate aeration and water transmission in the soil. Various inorganic cements, such as irreversibly dehydrated colloids of iron and aluminum oxides (Lutz, 1937; McIntyre, 1956; Deshpande *et al.*, 1964) as well as calcium carbonate (Kroth and Page, 1947), and especially stable organic resins ("humus") resulting from the decomposition of plant and animal residues (Swaby, 1950; Evans and Russell, 1959; Greenland *et al.*, 1962; Greenland, 1965) impart stability to soil aggregates.

When the clay is deflocculated, as under the influence of exchangeable sodium, the soil aggregates generally collapse. Aggregates are also vulnerable to the effects of water (through such phenomena as swelling and shrinkage, ice formation, the beating action of raindrops, and the scouring action of runoff). Excessive tillage and compaction also cause the breakdown of aggregates. On the other hand, close-growing perennial plants with extensive root systems, such as grasses, and certain types of microbial activity, promote soil aggregation. Synthetic "soil conditioners" are available which, when added to the soil in relatively small quantities, can cement and stabilize aggregates (Martin *et al.*, 1952; Hagin and Bodman, 1954; Haise *et al.*, 1955; Quastel, 1954).

Particularly vulnerable to structural deterioration and aggregate break-down is the soil surface zone, which in an agricultural field is often tilled, exposed to raindrop action (McIntyre, 1958), and compacted by traffic. Such deterioration often results in the formation of a dense and relatively impervious surface crust (Hillel, 1960), which in turn obstructs infiltration of water and free exchange of gas, and impedes seedling emergence (Hanks, 1960; Phillips and Kirkham, 1962). Thus, the development and maintenance of a desirable soil structure, optimal for crop growth, is a perpetual require-ment in agricultural soil management.

I. Summary

Soil physics deals with the physical properties and processes of the soil. The three phases or primary components of the soil are the solid particles, water, and air. Their relative quantities in the soil differ from place to place and from time to time. Their quantitative relationships can be characterized in terms of their mass or volume ratios. A simple diagrammatic scheme can help to define such useful concepts and indexes as density, porosity, void ratio, water content, etc. The main attributes of the solid phase constituting the matrix of soil are: texture (particularly clay content), specific surface area, and structure. The first attributes are constant and characteristic for a particular soil, while soil structure is a variable property which is difficult to measure and even more difficult to control in practice.

2 Physical Properties of Water

A. General

Water is the most prevalent substance on the surface of the earth, covering more than two-thirds of it. Even so-called dry land is frequently charged with and shaped by water. In the vapor form, it is always present in the atmosphere, even in the driest of climates. Water is the very stuff of life, the principal constituent of plants and animals. Yet, despite its ubiquity, water remains somewhat of an enigma, a substance of unique and in part unexplained attributes.

Knowledge of the basic physical properties of water is essential for an understanding of its behavior and function in nature, its interactions with the soil, and its state and movement in the soil–plant–atmosphere system as a whole.

B. Molecular Structure

One cubic centimeter of liquid water contains about 3.4×10^{22} (34,000 billion billion) molecules, the diameter of which is about 3 Å (3×10^{-8} cm). The chemical formula of water is H_2O, which signifies that each molecule consists of two atoms of hydrogen and one atom of oxygen. There are three isotopes of hydrogen (^1H, ^2H, ^3H), as well as three isotopes of oxygen (^{16}O, ^{17}O, ^{18}O), which can form 18 different combinations. However, all isotopes but ^1H and ^{16}O are quite rare.

The hydrogen atom consists of a positively charged proton and a negatively charged electron. The oxygen atom consists of a nucleus having the positive

29

charge of eight protons, and eight electrons, of which six are in the outer shell. Since the outer electron shell of the hydrogen lacks one electron and that of the oxygen lacks two electrons, one atom of oxygen can combine with two atoms of hydrogen in an electron-sharing molecule. The two hydrogen atoms link to the oxygen atom at an angle of about 105 deg from each other (as shown in Fig. 2.1) which is close to the tetrahedral angle of 109 deg (Pauling, 1960).

The asymmetrical arrangement of the hydrogens causes an imbalance of the electrostatic charges in the water molecule. On one side of the molecule

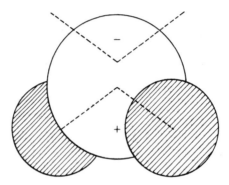

Fig. 2.1. Model of a water molecule consisting of one oxygen atom (white) and two hydrogen atoms (darkened).

(the oxygen side), an excess negative charge prevails, while on the other side (the hydrogen side), there is an excess positive charge. This distribution of charges creates an electrical *dipole*, or a *polarity*, which in turn imparts to water molecules an attraction for their neighbors at a certain orientation, and is also the reason why water is so good a solvent and why it adsorbs readily upon solid surfaces and hydrates ions and colloids.

Every hydrogen proton, while it is attached primarily to a certain molecule, is also attracted to the oxygen of the neighboring molecule, with which it forms a secondary link known as a *hydrogen bond*. This intermolecular link resulting from dipole attraction is not as strong as the primary attachment of the hydrogen to the oxygen of its own molecule, an attachment which results from the internal charge balance of the molecule. Accordingly, water can be regarded as a polymer of hydrogen-bonded molecules. This structure is most characteristically complete in ice crystals, in which each molecule is linked to four neighbors by means of four hydrogen bonds, thus forming a hexagonal lattice that is a rather open structure (Fig. 2.2). When the ice melts, this rigid structure collapses partially so that additional molecules

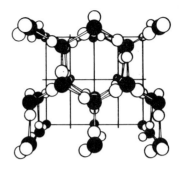

Fig. 2.2. Schematic structure of an ice crystal. The oxygen atoms are shown in black and the hydrogen atoms in white. The pegs linking adjacent molecules represent hydrogen bonds (after Buswell and Rodebush, 1956).

can enter into the intermolecular spaces and each molecule thus can have more than four near neighbors. For this reason, liquid water can be more dense than ice at the same temperature.

The orderly structure of ice does not totally disappear in the liquid state, as the molecules do not become entirely independent of each other. The polarity and hydrogen bonds continue to impart a crystal-like structure to liquid water (Morgan and Warren, 1938), except that this structure is not as rigid and permanent, but more flexible and transitory. Liquid molecules possess greater potential energy, having absorbed about 80 cal/gm in transition from the solid to the liquid state.

Although a number of different models of the liquid structure of water have been proposed in an attempt to account for the thermodynamic, spectroscopic, and transport properties of water, no single model explains all of the known properties of the liquid. Considerable evidence (Kavanu, 1964) supports the idea that hydrogen bonds in liquid water form an extensive three-dimensional network, the detailed features of which are probably shortlived. There are eight or nine known forms of ice, each stable over certain temperature ranges. There are also hydrated crystals in which the water structure might be described as icelike. In all of these forms, each water molecule is hydrogen-bonded to four others.

The question of whether hydrogen bonds are "broken" when ordinary ice melts, or are merely stretched and bent (as in the high-pressure transitions from one ice form to another), is a subject of controversy. "Continuum" models describe water as a hydrogen-bonded network with a continuous distribution of bond energies and geometries, while "mixture" models assume that an equilibrium exists between discrete molecular groupings with different numbers of hydrogen bonds per molecule (Narten and Levy, 1969).

According to the "flickering cluster" model (Frank and Wen, 1957; Nemethy and Scheraga, 1962), the molecules in liquid water associate and disassociate repeatedly in transitory or flickering clusters which have a crystalline internal structure. These microcrystals, as it were, form and melt so rapidly and randomly that, on a macroscopic scale, liquid water appears to behave as a homogeneous liquid. The cluster is visualized as short-lived (10^{-10} to 10^{-11} sec) and is continuously exchanging molecules with the adjacent unstructured phase (Fig. 2.3). However, this model was found difficult to reconcile with x-ray scattering data (Narten and Levy, 1969).

The strength of the hydrogen bonding and of the internal structure of water accounts for the fact that water, although of rather low molecular

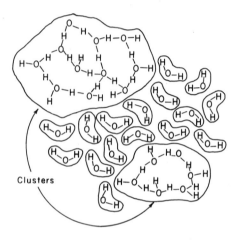

Fig. 2.3. Schematic illustration of "flickering clusters" of water microcrystals and free molecules in liquid water (after Nemethy and Scheraga, 1962).

weight, is a liquid and not a gas at normal temperatures. Furthermore, it is the hydrogen-bond effect which imparts to water unusually high values of specific heat and viscosity.

In particular, the specific heat capacity of liquid water is outstandingly high, being equal to 1 cal deg^{-1}, gm^{-1} at 15°C. By way of contrast, the specific heat of ice is about 0.5, of aluminum 0.215, of iron 0.106, of mercury only 0.033, and of air 0.17 cal/gm. The specific heat of dry soils is about 0.2 cal/gm, or about 0.15 cal/cm³. The specific heat of a water-saturated soil, on the other hand, can be more than twice as great.

Water, being a polar liquid, is strongly attracted to ions and electrostatically charged colloids. The formation of a hydration layer around ions or micelles can modify the internal structure of water. The water molecules thus bound lose energy, and the heat released is known as the *heat of solution* or, as in the case of clay particles, the *heat of wetting*.

C. Change of State

In transition from the solid to the liquid, and from the liquid to the gaseous state, hydrogen bonds must be disrupted (while in condensation and freezing, they must be reestablished). Hence, it requires relatively high temperatures and energy values to achieve these transitions. To thaw 1 gm of ice, 80 cal must be supplied; and conversely, the same energy (the latent heat of fusion) is released in freezing.

At the boiling point (100°C at atmospheric pressure), water passes from the liquid to the gaseous state and in so doing it absorbs 540 cal/gm. This amount of heat, known as the latent heat of vaporization, destroys the intermolecular structure and separates the molecules. Water can be vaporized at temperatures below 100°C, but such vaporization requires greater heat. At 25°C, for instance, the latent heat is 580 cal/gm. Sublimation is the direct transition from the solid state to vapor, and the heat required to effect it is equal to the sum of the latent heats of fusion and of vaporization.

D. Density and Compressibility

The open packing of water molecules in ice and liquid water accounts for their relatively low density. If the molecules were close-packed, the hypothetical density of water would be nearly 2 gm/cm^3, i.e., almost doubled. Unlike most substances, water exhibits a point of maximum density (at 4°C) below which the substance expands due to the formation of the hexagonal lattice structure, and above which the expansion is due to the increasing thermal motion of the molecules. The coefficient of thermal expansion of water is rather low, and in the normal temperature range of, say, 4°–50°C, the density decreases only slightly from 1.000 to 0.988 gm/cm^3. This change is generally considered negligible.

The compressibility of water c_w can be defined as the relative change in density with change in pressure:

$$c_w = \frac{1}{\rho_w} \frac{\partial \rho_w}{\partial P} \tag{2.1}$$

At 20°C and at atmospheric pressure, the compressibility of pure water is about 4.6×10^{-11} $cm^2/dyne$. In soil–water relationships, water can usually be taken to be incompressible.

E. Vapor Pressure

According to the kinetic theory, molecules in a liquid are in constant motion, which is an expression of their thermal energy. These molecules collide frequently, and occasionally one or another of them absorbs sufficient momentum to leap out of the liquid and into the atmosphere above it. Such a molecule, by virtue of its kinetic energy in the liquid phase, thus changes from the liquid to the gaseous phase. This kinetic energy is then lost in overcoming the potential energy of intermolecular attraction while escaping from the liquid. At the same time, some of the molecules in the gaseous phase may strike the surface of the liquid and be absorbed in it.

The relative rates of these two directions of movement depends upon the concentration of vapor in the atmosphere relative to its concentration at a state of equilibrium (i.e., when the movement in both directions is equal). An atmosphere that is at equilibrium with free and pure water is considered to be saturated with water vapor and the partial pressure of the vapor in such an atmosphere is called the *saturation (or equilibrium) vapor pressure.* The vapor pressure at equilibrium with any body of water depends upon the physical condition of the water (pressure and temperature) and its chemical condition (solutes) but does not depend upon the absolute or relative quantity of liquid or gas in the system.

The saturation vapor pressure increases with increasing temperature. As the kinetic energy of the molecules in the liquid increases, the evaporation rate increases and a higher concentration of vapor in the atmosphere is required for the rate of return to the liquid to match the rate of escape from it. A liquid arrives at its boiling point when the vapor pressure becomes equal to the atmospheric pressure. If the temperature range is not too wide, the dependence of saturation vapor pressure on temperature is expressible by the equation[1]

$$\ln p_0 = a - \frac{b}{T} \tag{2.2}$$

where $\ln p_0$ is the logarithm to the base e of the saturation vapor pressure p_0, T is the absolute temperature, and a and b are constants.

As mentioned earlier, the vapor pressure depends also upon the pressure of the liquid water. At equilibrium with drops of water which have a hydrostatic pressure greater than atmospheric, the vapor pressure will be greater than in a state of equilibrium with free water, which has a flat interface with

---

[1] This is a simplified version of the Clausius–Clapeyron equation $dp/dT = \Delta H_v / T(\bar{v}_v - \bar{v}_l)$, where ΔH_v is the latent heat of vaporization, and $\bar{v}_v$, $\bar{v}_l$ are the specific volumes of the vapor and liquid, respectively. If the vapor behaves as an ideal gas, $\bar{v}_v = RT/p$, hence $\ln/p = \Delta H_v/RT + \text{const.}$ (See, for instance, Barrow, 1961).

the atmosphere. On the other hand, in equilibrium with adsorbed or capillary water under a hydrostatic pressure smaller than atmospheric, the vapor pressure will be smaller than that in equilibrium with free water. The curvature of drops is considered to be positive, as these drops are convex toward the atmosphere, whereas the curvature of capillary water menisci is considered negative, as they are concave toward the atmosphere.[2]

Water present in the soil invariably contains solutes, mainly electrolytic salts, in highly variable concentrations. Thus, soil water should properly be called the *soil solution*. The composition and concentration of the soil solution affect soil behavior. While in humid regions, the soil solution may have a concentration of but a few parts per million, in arid regions the concentration may become as high as several per cent. The ions commonly present are H^+, Ca^{2+}, Mg^{2+}, Na^+, NH_4^+, OH^-, Cl^-, HCO_3^-, NO_3^-, SO_4^{2-}, and CO_3^{2-}. Since the vapor pressure of electrolytic solutions is lower than that of pure water, soil water also has a lower vapor pressure, even when the soil is saturated. In an unsaturated soil, the capillary and adsorptive effects further lower the potential and the vapor pressure, as will be shown in the next chapter.

Vapor pressure can be expressed in units of dynes per square centimeter, or bars, or millimeters of mercury, or in other convenient pressure units. The vapor content of the atmosphere can also be expressed in units of *relative humidity* (the ratio of the existing vapor pressure to the saturation vapor pressure at the same temperature), *vapor density* (the mass of water vapor per unit volume of the air), the *specific humidity* of the air (the mass of water vapor per unit mass of the air), the *saturation (or vapor pressure) deficit* (the difference between the existing vapor pressure and the saturation vapor pressure at the same temperature), and the *dew-point temperature*
The presence of solutes generally lowers the vapor pressure.[3]

---

[2] For water in capillaries, in which the air–water interface is concave, the Kelvin equation applies:

$$-(\mu_l - \mu_l^\circ) = RT \ln(p_l^\circ/p_l) = 2\gamma \bar{v}_l \cos a/r_c$$

in which $(\mu_l - \mu_l^\circ)$ is the change in potential of the water due to the curvature of the air–water interface, γ is the surface tension of water, a the contact angle, $\bar{v}_l$ the partial molar volume of water, and r_c the radius of the capillary.

The concept of water potential will be elucidated more fully in Chapter 3.

[3] The equation is:

$$\pi_0 \bar{v}_l = RT \ln(p_l^\circ/p_l) = \mu_l - \mu_l^\circ,$$

where π_0 is the osmotic pressure of a nonvolatile solute, μ_l° and p_l° are the chemical potential and vapor pressure of the liquid in its standard state, and μ_l and p_l are the same for the solution.

(the temperature at which the existing vapor pressure becomes equal to the saturation vapor pressure, i.e., the temperature at which a cooling body of air with a certain vapor content will begin to condense dew).

F. Surface Tension

Surface tension is a phenomenon occurring typically at the interface of a liquid and a gas. The liquid behaves as if it were covered by an elastic membrane in a constant state of tension which tends to cause the surface to contract. If we draw an arbitrary line of length L on a liquid surface, there will be a force F pulling the surface to the right of the line and an equal force pulling the surface leftwards. The ratio F/L is the surface tension and its

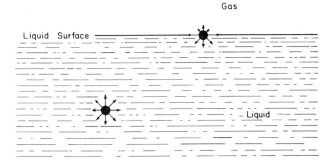

Fig. 2.4. Cohesive forces acting on a molecule inside the liquid and at its surface.

dimensions are those of force per unit length (dynes per centimeter, or grams per second-squared). The same phenomenon can also be described in terms of energy. Increasing the surface area of a liquid requires the investment of energy, which remains stored in the enlarged surface, just as energy can be stored in a stretched spring, and it can perform work if the enlarged surface is allowed to contract again. Energy per unit area has the same dimensions as force per unit length (ergs per square centimeter or grams per second-squared).

An explanation for occurrence of surface tension is given in Fig. 2.4. Molecule A inside the liquid is attracted in all directions by equal cohesive forces, while molecule B at the surface of the liquid is attracted into the denser liquid phase by a force greater than the force attracting it into the gaseous phase. This unbalanced force draws the surface molecules inward into the liquid and results in the tendency for the surface to contract.

As we shall see later, surface tension is associated with the phenomenon of capillarity. When the interface of the liquid and the gas is not planar but curved (concave or convex), a pressure difference between the two phases is indicated, since the surface-tension forces have a resultant normal to the surface, which, in equilibrium, must be counteracted by a pressure difference across the interface. If we stretch a rubber membrane as a boundary between two air cells of different pressure, this membrane will bulge into the side having the lower pressure. Similarly, a liquid with an interface which is convex toward the atmosphere is under a pressure greater than atmospheric; a liquid with an interface concave toward the gaseous phase is at a pressure smaller than atmospheric, and a liquid with a flat interface is at the same pressure as the atmosphere.[4]

Different liquids differ in their surface tension, as illustrated in the following list:

Water, 72.7 dynes/cm (at 20°C)
Ethyl ether, 17 dynes/cm
Ethyl alcohol, 22 dynes/cm
Benzene, 29 dynes/cm
Mercury, 430 dynes/cm.

Surface tension also depends upon temperature, generally decreasing almost linearly as the temperature rises. Thermal expansion tends to decrease the density of the liquid, and therefore to reduce the cohesive forces at the surface as well as inside the liquid phase. The decrease of surface tension is accompanied by an increase in vapor pressure.

Soluble substances can influence surface tension in either direction. If the affinity of the solute molecules or ions to water molecules is greater than the affinity of the water molecules to one another, then the solute tends to be drawn into the solution and to cause an increase in the surface tension (e.g., electrolytes).[5] If, on the other hand, the cohesive attraction between water molecules is greater than their attraction to the solute molecules, then the latter tend to be relegated or concentrated more toward the surface, reducing its tension (e.g., many organic solutes, particularly detergents).

---

[4] An important difference between a rubber membrane and a liquid surface is that the former increases its tension as it is stretched and reduces its tension as it is allowed to contract, while the liquid surface retains a constant surface tension regardless of curvature.

[5] E.g., a 1.0% NaCl concentration increases the surface tension by 0.17 dynes/cm at 20°C.

Table 2.1

PHYSICAL PROPERTIES OF LIQUID WATER

Temperature °C	Density (gm/cm³)	Specific heat (cal/gm-deg)	Latent heat (vaporization) (cal/gm)	Surface tension (gm/sec²)	Thermal conductivity (cal/cm-sec-deg) × 10⁻³	Viscosity (gm/cm-sec) × 10⁻²	Kinematic viscosity (cm²/sec)
−10	0.99794	1.02	603.0	—	—	—	—
− 5	0.99918	1.01	600.0	76.4	—	—	—
0	0.99987	1.007	597.3	75.6	1.34	1.787	0.0179
4	1.00000	1.005	595.1	75.0	1.36	1.567	0.0157
5	0.99999	1.004	594.5	74.8	1.37	1.519	0.0152
10	0.99973	1.001	591.7	74.2	1.40	1.307	0.0131
15	0.99913	1.000	588.9	73.4	1.42	1.139	0.0114
20	0.99823	0.999	586.0	72.7	1.44	1.002	0.01007
25	0.99708	0.998	583.2	71.9	1.46	0.890	0.00897
30	0.99568	0.998	580.4	71.1	1.48	0.798	0.00804
35	0.99406	0.998	577.6	70.3	1.50	0.719	0.00733
40	0.99225	0.998	574.7	69.5	1.51	0.653	0.00661
45	0.99024	0.998	571.9	68.7	1.53	0.596	0.00609
50	0.98807	0.999	569.0	67.9	1.54	0.547	0.00556

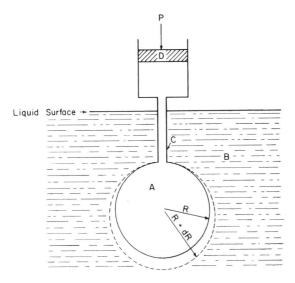

Fig. 2.5. A trial illustrating the relation between surface tension, radius of curvature, and bubble pressure.

G. Curvature of Water Surfaces and Hydrostatic Pressure

In order to illustrate the relationship between surface curvature and pressure, we shall carry out a hypothetical experiment, as illustrated in Fig. 2.5. This figure shows a bubble of gas A, blown into a liquid B through a capillary C. If we neglect the influence of gravitation and the special conditions occurring at the edge of the tube, we can expect the bubble to be spherical (a shape that is obtained because it affords the smallest surface area for a given volume), with a radius R.

If we now add a small amount of gas by lowering the piston D under a pressure greater than atmospheric by a magnitude P, the radius of the bubble will increase to $R + dR$. This will in turn increase the surface area of the bubble by $4\pi(R + dR)^2 - 4\pi R^2 = 8\pi R \ dR$ (neglecting the second-order differential terms). Increasing the surface area of the bubble required the investment of work against the surface tension γ, and the amount of this work is $\gamma 8\pi R \ dR$. Simultaneously, we have increased the volume of the bubble by $\frac{4}{3}\pi(R + dR)^3 - \frac{4}{3}\pi R^3 = 4\pi R^2 \ dR$. This increase in volume against the pressure P involved work in the amount $P4\pi R^2 \ dR$. The two expressions for the quantity of work performed must be equal, i.e., $\gamma 8\pi R \ dR = P4\pi R^2 \ dR$. Therefore,

$$P = \frac{2\gamma}{R} \qquad (2.3)$$

This important equation shows that the difference P between the pressure of the bubble and the pressure of the water surrounding it is directly proportional to the surface tension and inversely proportional to the radius of the bubble; thus, the smaller the bubble is, the greater is its pressure.[6]

If the bubble is not spherical, then instead of Eq. (2.6) we will obtain

$$P = \gamma \left(\frac{1}{R_1} + \frac{1}{R_2} \right) \tag{2.4}$$

where R_1 and R_2 are the principal radii of curvature for a given point on the interface. This equation reduces to the previous one whenever $R_1 = R_2$.

H. Contact Angle of Water on Solid Surfaces

If we place a drop of liquid upon a dry solid surface, the liquid will usually displace the gas which covered the surface of the solid and spread

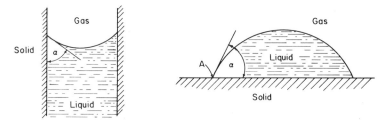

Fig. 2.6. The contact angle of a meniscus in a capillary tube and of a drop resting upon a plane solid surface.

over that surface to a certain extent. Where its spreading will cease and the edge of the drop will come to rest, its interface with the gas will form a typical angle with its interface with the solid. This angle, termed the *contact angle*, is illustrated in Fig. 2.6.

Viewed two-dimensionally on a cross-sectional plane, the three phases meet at a point A and form three angles with the sum of 360 deg. If we assume the angle in the solid to be 180 deg, and if we designate the angle in the liquid as α, the angle in the gaseous phase will be $180 - \alpha$ (deg).

We can perhaps simplify the matter by stating that, if the adhesive forces between the solid and liquid are greater than the cohesive forces inside the liquid itself, and greater than the forces of attraction between the gas and solid, then the solid–liquid contact angle will tend to be acute and the liquid

---

[6] The reader is invited to ponder the fact that, in the experiment described in Fig. 2.5, blowing additional air into the bubble by lowering the piston results in a decrease, not an increase, of internal pressure of the bubble.

will wet the solid. A contact angle of zero would mean the complete flattening of the drop and perfect wetting of the solid surface by the liquid. It would be as though the solid surface had an absolute preference for the liquid over the gas. A contact angle of 180 deg (if it were possible) would mean a complete nonwetting or rejection of the liquid by the gas-covered solid, i.e., the drop would retain its spherical shape without spreading over the surface at all (assuming no gravity effect).

In order for a drop resting on a solid surface to be in equilibrium with that surface and with a gas phase, the vector sum of the three forces arising from

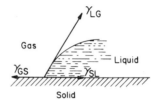

Fig. 2.7. Equilibrium of surface-tension forces at the edge of a drop.

the three types of surface tension present must be zero. On the solid surface drawn in Fig. 2.7, the sum of the forces pulling leftward at the edge of the drop must equal the sum of the forces pulling to the right:

$$\gamma_{gs} = \gamma_{sl} + \gamma_{lg} \cos \alpha$$

and therefore

$$\cos \alpha = \frac{\gamma_{gs} - \gamma_{sl}}{\gamma_{lg}} \qquad (2.5)$$

where γ_{sl} is surface tension between solid and the liquid, γ_{gs} is surface tension between gas and solid, and γ_{lg} is the surface tension between liquid and gas. Each of these surface tensions tends to decrease its own interface. Reducing the interfacial tensions γ_{lg} and γ_{sl} (as with the aid of a detergent) can increase $\cos \alpha$ and decrease the contact angle α, thus promoting the wetting of the solid surface by the liquid.

The contact angle of a given liquid on a given solid is generally constant under given physical conditions. This angle, however, can be different in the case of a liquid that is advancing upon the solids ("wetting angle" or "advancing angle") than for a liquid that is receding upon the solid surface (the "retreating" or "receding" angle). The wetting angle of pure water upon clean and smooth inorganic surfaces is generally zero, but where the surface is rough or coated with adsorbed surfactants of a hydrophobic nature, the contact angle, and especially the wetting angle, can be considerably greater than zero.

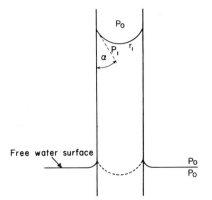

Fig. 2.8. Capillary rise.

I. Capillarity

A capillary tube dipped in a body of water will form a meniscus as the result of the contact angle of water with the walls of the tube. The curvature of this meniscus will be greater (i.e., the radius of curvature smaller) the narrower the tube. The occurrence of curvature causes a pressure difference to develop across the liquid–gas interface. A liquid with an acute contact angle (e.g., water on glass) will form a meniscus concave toward the air, and therefore the liquid pressure under the meniscus P_1 will be smaller than the atmospheric pressure P_0 (Fig. 2.8). For this reason, the water inside the tube, and the meniscus, will be driven up the tube from its initial location (shown as a dashed curve in Fig. 2.8) by the greater pressure of the free water[7] outside the tube at the same level, until the pressure difference between the water inside the tube and the water under the flat surface outside the tube is relieved by the counterhydrostatic pressure of the water column in the capillary tube.

In a cylindrical capillary tube, the meniscus assumes a spherical shape. When the contact angle of the liquid on the walls of the tube is zero, the meniscus is a hemisphere (and in two-dimensional drawing can be represented as a semicircle) with its radius of curvature equal to the radius of the capillary tube. If, on the other hand, the liquid contacts the tube at an angle greater than zero but smaller than 90%, then the diameter of the tube ($2r$) is the length of a chord cutting a section of a circle with an angle of $\pi - 2\alpha$, as shown in Fig. 2.9. Thus,

$$R = \frac{r}{\cos \alpha} \qquad (2.6)$$

[7] By "free water," we refer to water at atmospheric pressure, under a horizontal air–water interface. This is in contrast with water that is constrained by capillarity or adsorption, and is at an equivalent pressure smaller than atmospheric (i.e., tension).

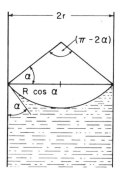

Fig. 2.9. The geometric relationship of the radius of curvature R to the radius of the capillary r and the contact angle α.

where R is the radius of curvature of the meniscus, r the radius of the capillary, and α the contact angle.

The pressure difference P between the capillary water (under the meniscus) and the atmosphere, therefore, is

$$P = \frac{2\gamma \cos \alpha}{r} \tag{2.7}$$

and the height of capillary rise h is

$$h = \frac{2\gamma \cos \alpha}{g(\rho_l - \rho_g)r} \tag{2.8}$$

where ρ_g is the density of the gas (which is generally neglected), ρ_l the density of the liquid, g the acceleration of gravity, r the capillary radius, α the contact angle, and γ the surface tension between the liquid and the air.

When the liquid surface is concave, the center of curvature lies outside the liquid and the curvature, by convention, is regarded as negative. Thus, for a concave meniscus such as that of water in a clean glass capillary, P is negative with reference to the atmosphere, indicating a capillary pressure deficit (or subpressure). For a convex meniscus (such as that of mercury in glass, or of water in an oily or otherwise water-repellent tube), P is positive and capillary depression, rather than capillary rise, will result.

J. Adsorption of Water on Solid Surfaces

Adsorption is another type of interfacial phenomenon resulting from the differential forces of attraction or repulsion occurring among molecules of different phases at their contact surfaces. As a result of these cohesive and adhesive forces, the contact zone may exhibit a concentration or a density of material different from that inside the phases themselves. According to the different phases which may come into contact, various types of adsorption can occur, such as the adsorption of gases upon solids, of gases upon liquid surfaces, of liquids upon solids, etc. In some cases, it is possible to distinguish between *chemical adsorption* and *physical adsorption*, with the former representing an irreversible chemical interaction between the adsorbed and adsorbing phases. However, this distinction is often arbitrary.

The interfacial forces of attraction or repulsion may themselves be of different types, including electrostatic or ionic (Coulomb) forces, intermolecular forces known as van der Waals' and London forces, and short-range repulsive (Born) forces. The adsorption of water upon solid surfaces is generally of an electrostatic nature. The polar water molecules attach to the charged faces of the solids. The adsorption of water is the mechanism causing the strong retention of water by clay soils at high suctions.

The interaction of the charges of the solid with the polar water molecules may impart to the adsorbed water a distinct and rigid structure in which the water dipoles assume an orientation dictated by the charge sites on the solids. Some investigators believe that the adsorbed layer or "phase" has a quasi-crystalline, icelike structure and can assume a thickness of 10–20 Å (i.e., 3–7 molecular layers) or more (Low, 1961). This adsorbed water layer may have mechanical properties of strength and viscosity which differ from those of ordinary liquid water at the same temperature.

The adsorption of water upon clay surfaces is an exothermic process, resulting in the liberation of an amount of heat known as the *heat of wetting*. Anderson (1926) found a linear relationship between heat of wetting and exchange capacity. Janert (1934) traced the relationship between the heat of wetting and the nature of the exchange cations. The distinction

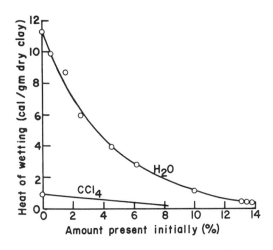

Fig. 2.10. Heat of wetting in relation to initial liquid content (after Janert, 1934).

between polar and nonpolar adsorption is illustrated in Fig. 2.10, in which water and carbon tetrachloride are compared for a "brick clay."

K. Osmotic Pressure

Osmotic pressure is a property of solutions, expressing the decrease of the potential energy of water in solution relative to that of pure water. When an aqueous solution is separated from pure water (or from a solution of lower concentration) by a membrane that is permeable to water alone (i.e., a "selectively permeable" or "semipermeable" membrane), water will tend to diffuse, or osmose, through the membrane into the more concentrated solution, thus diluting it and reducing the potential-energy difference across the membrane. The osmotic pressure is the counter pressure which must be applied to the solution to prevent the osmosis of water into it (Fig. 2.11).

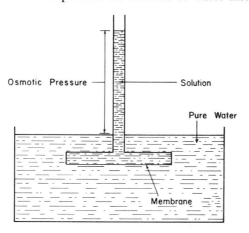

Fig. 2.11. Schematic illustration of an osmometer.

In dilute solutions, the osmotic pressure is generally proportional to the concentration of the solution and to its temperature according to the following equation:

$$P_s = kTC_s \qquad (2.9)$$

where P_s is the osmotic pressure, T the absolute temperature, and C_s the concentration of the solute.[8]

[8] The osmotic pressure increase with temperature is associated with the corresponding increase of the molecular diffusivity (self-diffusion coefficient) of water, D_w. According to the Einstein–Stokes equation:

$$D_w = kT/6\pi r \nu$$

where $k = R/N$, the Boltzmann constant (1.37×10^{-6} erg $°K$); r is the rotation radius of the molecule (~ 1.5 Å), and ν is the viscosity.

Table 2.2

PHYSICAL PROPERTIES OF WATER VAPOR

Temperature (C°)	Saturation vapor pressure (Torr)		Saturation vapor density (gm/cm³)		Diffusion coefficient (cm²/sec)
	Over liquid	Over ice	Over liquid	Over ice	
−10	2.15	1.95	2.36	2.14	0.211
−5	3.16	3.01	3.41	3.25	—
0	4.58	4.58	4.85	4.85	0.226
5	6.53	—	6.80	—	—
10	9.20	—	9.40	—	0.241
15	12.78	—	12.85	—	—
20	17.52	—	17.30	—	0.257
25	23.75	—	23.05	—	—
30	31.82	—	30.38	—	0.273
35	42.20	—	39.63	—	—
40	55.30	—	51.1	—	0.289
45	71.90	—	65.6	—	—
50	92.50	—	83.2	—	—

Table 2.3

RELATION OF PRESSURE (OR TENSION) OF WATER UNDER CURVED SURFACES TO
VAPOR PRESSURE

Radius of curvature (cm)	Hydrostatic pressure (bars)	Height of capillary rise (cm)	Relative vapor pressure at 15°C
10^{-6}	1.5×10^2	—	1.114
10^{-5}	1.5×10	—	1.011
10^{-4}	1.5	—	1.001
10^{-3}	1.5×10^{-1}	—	1.0001
—	0	0	1.0000
-10^{-1}	-1.5×10^{-3}	1.5	1.0000
-10^{-2}	-1.5×10^{-2}	1.5×10	1.0000
-10^{-3}	-1.5×10^{-1}	1.5×10^2	1.0000
-10^{-4}	-1.5	1.5×10^3	0.9989
-10^{-5}	-1.5×10	1.5×10^4	0.9890
-10^{-6}	-1.5×10^2	1.5×10^5	0.8954
-10^{-7}	-1.5×10^3	1.5×10^6	0.3305
-10^{-8}	-1.5×10^4	1.5×10^7	0.000016

An increase in osmotic pressure of a solution is usually accompanied by a decrease in the vapor pressure, a rise of the boiling point, and a depression of the freezing point.

L. Solubility of Gases

The concentration of gases in water generally increases with pressure and decreases with temperature. According to Henry's law, the mass concentration of the dissolved gas C_m is proportional to the partial pressure of the gas p_i:

$$C_m = s_c \frac{p_i}{p_0} \qquad (2.10)$$

where s_c is the solubility coefficient of the gas in water and p_0 is the total pressure of the atmosphere. The volume concentration is similarly proportional:

$$C_v = s_v \frac{p_i}{p_0} \qquad (2.11)$$

where s_v is the solubility expressed in terms of volume ratios (i.e., C_v = volume of dissolved gas relative to the volume of water).

Table 2.4 gives the values of s_v for several atmospheric gases at various temperatures.

Table 2.4.

SOLUBILITY COEFFICIENTS OF GASES IN WATER

Temperature (°C)	Nitrogen (N_2)	Oxygen (O_2)	Carbon dioxide (CO_2)	Air (without CO_2)
0	0.0235	0.0489	1.713	0.0292
10	0.0186	0.0380	1.194	0.0228
20	0.0154	0.0310	0.878	0.0187
30	0.0134	0.0261	0.665	0.0156
40	0.0118	0.0231	0.530	—

M. Viscosity

When a fluid is moved in shear (that is to say, when adjacent layers of fluid are made to slide over each other), the force required is proportional to the velocity of shear. The proportionality factor is called the *viscosity*.

As such, it is the property of the fluid to resist the rate of shearing, and can be visualized as an internal friction. The coefficient of viscosity v is defined as the force per unit area necessary to maintain a velocity difference of 1 cm/sec between two parallel layers of fluid which are 1 cm apart.

The viscosity equation is

$$\tau = \frac{F_s}{A} = v\frac{du}{dx}$$

where τ is the shearing stress, consisting of a force F_s acting on an area A; v (dimensions: mass/length $\times$ time) is the coefficient of viscosity; and du/dx is the velocity gradient perpendicular to the stressed area A.

The ratio of the viscosity to the density of the fluid is called the kinematic viscosity, designated v_k. It expresses the shearing-rate resistance of a fluid mass independently of the density. Thus, while the viscosity of water exceeds that of air by a factor of about 50 (at room temperature), its kinematic viscosity is actually lower.

Fluids of lower viscosity flow more readily and are said to be of greater *fluidity* (which is the reciprocal of viscosity). As shown in Table 2.1, the viscosity of water decreases by about 3 % per 1°C rise in temperature, and thus decreases by half as the temperature increases from 5° to 35°C. The viscosity is also affected by type and concentration of solutes.

N. Summary

Because of its internal molecular makeup, water is a polar substance and this fact influences its intermolecular structure in the solid and liquid phases, the nature of its transition from one phase to another, its cohesive and adhesive properties, its properties as a solvent, and many other phenomena. The vapor pressure of water characterizes its energy state. Additional attributes of water which influence its condition and behavior in the soil are surface tension, adsorption, osmotic pressure, and viscosity.

3 The State of Water in the Soil

A. Energy State of Soil Water

Soil water, like other bodies in nature, can contain energy in different quantities and forms. Classical physics recognizes two principal forms of energy: kinetic and potential. Since the movement of water in the soil is quite slow, its kinetic energy, which is proportional to the velocity squared, is generally considered to be negligible. On the other hand, the potential energy, which is due to position or internal condition, is of primary importance in determining the state and movement of water in the soil.

The potential energy of soil water varies over a very wide range. Differences in potential energy of water between one point and another give rise to the tendency of water to flow within the soil. The spontaneous and universal tendency of all matter in nature is to move from where the potential energy is higher to where it is lower, and for each parcel of matter to equilibrate with its surroundings. Soil water obeys the same universal pursuit of equilibrium. It moves constantly in the direction of decreasing potential energy. The rate of decrease of potential energy with distance is in fact the moving force causing flow. A knowledge of the relative energy state of soil water at each point within the soil can allow us to evaluate the forces acting on soil water in all directions, and to determine how far the water in a soil system is from equilibrium (i.e., a state of uniform potential energy throughout the system).

Clearly, therefore, it is not the absolute amount of potential energy "contained" in the water which is important in itself, but rather the relative level of that energy in different regions within the soil. The concept of

49

soil-water potential[1] is a criterion, or yardstick, for this energy. It expresses the specific potential energy of soil water relative to that of water in a standard reference state. The standard state generally used is that of a hypothetical reservoir of pure and free water, at atmospheric pressure, at the same temperature as that of soil water (or at any other specified temperature), and at a given and constant elevation. Since the elevation of this hypothetical reservoir can be set at will, it follows that the potential which is determined by comparison with this standard is not absolute, but by employing even so arbitrary a criterion we can determine the relative magnitude of the specific potential energy of water at different locations or times within the soil.

Just as an energy increment can be viewed as the product of a force by a distance increment, so the ratio of an energy to a distance increment can be viewed as constituting a force. Accordingly, the force acting on soil water, directed from a zone of higher to a zone of lower potential, is equal to the negative *potential gradient* $(-d\phi/dx)$, which is the change of energy potential ϕ with distance x. The negative sign indicates that the force acts in the direction of decreasing potential.

The concept of soil-water potential is of great fundamental importance. This concept replaces the arbitrary categorizations which prevailed in the early stages of the development of soil physics and which purported to recognize and classify different "forms" of soil water: e.g., "gravitational water," "capillary water," "hygroscopic water," etc. The fact is that all of soil water, not merely a part of it, is affected by the earth's gravitational field, so that in effect it is all "gravitational." Furthermore, the laws of capillarity do not begin or cease at certain values of wetness or pore sizes.

In what way, then, does water differ from place to place and from time to time within the soil? Not in "form," but in potential energy. The possible values of soil-water potential are continuous, and do not exhibit any abrupt discontinuities or changes from one condition to another (excepting perhaps changes in phase). Rather than attempt to classify soil water, the more valid approach is to characterize its potential-energy state.

When the soil is saturated and its water is at a hydrostatic pressure greater than the atmospheric pressure (as, for instance, under a water-table) the potential-energy level of that water may be greater than that of the "reference-state" reservoir described, and water will tend to move spontaneously from the soil into such a reservoir. If, on the other hand, the soil is moist but unsaturated, its water will no longer be free to flow out toward a reservoir at atmospheric pressure. On the contrary, the spontaneous tendency will be

---

[1] The potential concept was first applied to soil water by Buckingham, in his classic and still-pertinent paper on the "capillary" potential (1907). Gardner (1920) showed how this potential is dependent upon the water content. Richards (1931) developed the tensiometer for measuring it *in situ*.

for the soil to draw water from such a reservoir if placed in contact with it, much as a blotter draws ink.

Under hydrostatic pressures greater than atmospheric, the potential of soil water (in the absence of osmotic effects) is greater than that of the reference state and therefore can be considered "positive." In an unsaturated soil, the water is constrained by capillary and adsorptive forces, hence its energy potential is generally "negative," and its equivalent hydrostatic pressure is less than that of the reference state.

Under normal conditions in the field, the soil is generally unsaturated and the soil-water potential is negative. Its magnitude at any point depends not only on hydrostatic pressure but also upon such additional physical factors as elevation (relative to that of the reference elevation), concentration of solutes, and temperature.

B. Total Soil-Water Potential

We have already described the energy potential of soil water in a qualitative way. Thermodynamically, this energy potential can be regarded in terms of the difference in partial specific free energy between soil water and "standard" water. More explicitly, a soil physics terminology committee of the International Soil Science Society (Aslyng et al., 1963) defined the total potential of soil water as "the amount of work that must be done per unit quantity of pure water in order to transport reversibly and isothermally an infinitesimal quantity of water from a pool of pure water at a specified elevation at atmospheric pressure to the soil water (at the point under consideration)."

This is merely a formal definition, since in actual practice the potential is not measured by transporting water as per the definition, but by measuring some other property related to the potential in some known way (e.g., hydrostatic pressure, vapor pressure, elevation, etc.). The definition specifies transporting an "infinitesimal quantity," in any case, to ensure that the determination procedure does not change either the reference state (i.e., the pool of pure, free water) or the soil-water potential being measured. It should be recognized that this definition provides a conceptual rather than an actual working tool. It can be argued (in view of the hysteresis phenomenon to be discussed in Section I) that no change in soil wetness can in practice be carried out reversibly, or that the total potential need not be restricted to isothermal conditions. A most serious difficulty is encountered in attempting to allocate the total potential among the various components or mechanisms comprising it, since these may not be mutually independent.

The above definition is based upon the specific differential Gibbs-free-energy function. The differential form provides a criterion of equilibrium and of the direction in which changes can be expected to occur in non-equilibrium systems. Philip (1960) introduced the integral form of the thermo-dynamic function,[2] to provide a criterion of the total potential energy of a system during transition from one state to another.

Soil water is subject to a number of force fields, which cause its potential to differ from that of pure, free water. Such force fields result from the attraction of the solid matrix for water, as well as from the presence of solutes and the action of external gas pressure and gravitation. Accordingly, the *total potential* of soil water can be thought of as the sum of the separate contributions of these various factors, as follows:

$$\phi_t = \phi_g + \phi_p + \phi_o + \cdots \tag{3.1}$$

where ϕ_t is the *total potential*, ϕ_g the *gravitational potential*, ϕ_p the *pressure* (or *matric*) *potential*, ϕ_o the *osmotic potential*, and the dots on the right side signify that additional terms are theoretically possible.

Not all of the separate potentials given above act in the same way, and their separate gradients may not always be equally effective in causing flow (for example, the osmotic potential gradient requires a semipermeable membrane to induce liquid flow). The main advantage of the total-potential concept is that it provides a unified measure by which the state of water can be evaluated at any time and everywhere within the soil–plant–atmosphere continuum.

C. Thermodynamic Basis of the Potential Concept

It might be useful at this point to digress from our topic of soil water in order to clarify the thermodynamic background of the potential concept.

Over the past few decades, numerous attempts have been made to apply the principles and terminology of thermodymamics to the retention and movement of water in the soil–plant system. An early and comprehensive effort in this direction was made by Edlefsen and Anderson (1943). Classical thermodynamics (Guggenheim, 1959) deals with equilibrium states and reversible processes and can thus serve only to describe the forces acting on water and its energy of retention. However, equilibrium states occur only rarely in nature, and spontaneous processes tend to be irreversible. To describe such processes, a branch of thermodynamics known as non-equilibrium or irreversible thermodynamics, has been developed in recent years (Prigogine, 1961; de Groot, 1963; Katchalsky and Curran, 1965). The application of

---

[2] Defined for a small element of soil of a given wetness as minus the work required, isothermally and reversibly, per unit quantity of water, to *completely* remove the water from the soil and to transform it into pure, free water at some datum level.

irreversible thermodynamics to soil–water phenomena will be mentioned in Chapter 5, while the present chapter is based on classical thermodynamic relations.

The potential concept depends ultimately upon the first and second laws of classical thermodynamics. The first law is merely the well-known energy conservation law, which states that energy can be converted from one form to another, but can neither be created nor destroyed. In equation form:

$$dQ = dU + PdV + dW \qquad (3.2)$$

where dQ is heat added to the system, dU is change in internal energy U of the system, PdV is the work of expansion done by the system (P pressure, V volume), and dW is all the other work done by the system on its surroundings.

The second law of thermodynamics specifies that the direction of change in an isolated system is always toward equilibrium. This law has subtle and far-reaching implications. It has been stated verbally in several different ways,[3] none of which appears to convey its complete meaning to the layman. Mathematically, the second law rests on the definition of two properties: the *absolute temperature* T (being always positive), and the entropy S, such that

$$
\begin{aligned}
dQ &= T\,dS \text{ for reversible processes} \\
dQ &< T\,dS \text{ for irreversible processes}
\end{aligned}
\qquad (3.3)
$$

where dQ is, as before, the heat input into the system, and dS is the change in entropy. The intensive property[4] of temperature is familiar and hence intuitively understandable. The meaning of entropy, however, is not readily apparent. It is a measure of the internal disorder or randomness of a system. The change in entropy is equal to the ratio of the heat input to the temperature of the system, i.e., $dS = dQ/T$. In irreversible processes, dS in the system is greater than zero, thus the entropy tends to increase spontaneously.[5]

---

[3] For instance (Reid, 1960): "Heat can be conveyed from a lower to a higher temperature only by expenditure of work," or "It is impossible to devise a machine whose only net effect is to remove heat from a reservoir and to lift a weight," or "In the neighborhood of a given state there are other states which cannot be by adiabatic tranformation," or "All natural or spontaneous processes are irreversible," or "The total amount of entropy in nature is increasing".

[4] *Intensive properties* (e.g., temperature, pressure, concentration) are independent of the size of the system, whereas *extensive properties* (e.g., mass, volume) are defined by the system as a whole.

[5] Values for the entropy of water at different temperatures were reported by de Jong (1968):

Temperature (°K):	0	250	273	273	298	298
State:	—	solid	solid	liquid	liquid	vapor
Entropy (cal/mol/°K):	0	9.0	9.8	15.1	16.7	45.1

The second law of thermodynamics can now be stated as

$$dU = TdS - PdV \tag{3.4}$$

where dU is, again the change in internal energy U of the system.

In a system of variable composition, the total differential of the internal energy can be expressed as a function of S, V and n_i, where n_i is the number of moles of a given component in the system (Guggenheim, 1959; Slatyer, 1967):

$$dU = \left(\frac{\partial U}{\partial S}\right)_{V,n_i} dS + \left(\frac{\partial U}{\partial V}\right)_{S,n_i} dV + \left(\frac{\partial U}{\partial n_i}\right)_{S,V,n_j} dn_i \tag{3.5}$$

A very useful thermodynamic quantity is the *Gibbs free energy*, G, defined as

$$G = U + PV - TS \tag{3.6}$$

The *chemical potential*, μ_i, of a component in a system of variable composition is defined as the partial molal Gibbs free energy of that component, $\bar{G}_i$. The change in the free energy of the system with change in the concentration of the component considered is equivalent to

$$\bar{G}_i = \left(\frac{\partial G}{\partial n_i}\right) T,P,n_j = \mu_i \tag{3.7}$$

The total differential of the chemical potential is

$$d\mu_i = \left(\frac{\partial \mu_i}{\partial T}\right)_{P,n_i} dT - \left(\frac{\partial \mu_i}{\partial P}\right)_{T,n_i} dP - \left(\frac{\partial \mu_i}{\partial n_i}\right)_{T,P,n_j} dn_i \tag{3.8}$$

Thus the chemical potential is an expression of the potential energy state of a component in a mixed system in the absence of external forces, i.e., when temperature, pressure and composition are the only effective variables. The chemical potential thus excludes the effects of gravitational, centrifugal or electrical force fields. The chemical potential is constant in the system when the intensive parameters of temperature, pressure and concentration are constant and a state of equilibrium exists. In an unequilibrated system, a difference in chemical potential of a component between two locations determines the direction (but not the rate) in which the component will tend to move spontaneously within the system.

How these relationships apply in the case of soil water is still a matter of some controversy (Edelfsen and Anderson, 1943; Low, 1951; Low and Deming, 1953; Babcock and Overstreet, 1955, 1957; Taylor and Slatyer, 1960; Babcock, 1963). A comprehensive and critical review of this subject was published by Bolt and Frissel (1960). The major difficulty encountered in the formulation of a working equation for the potential of soil water is the selection of the independent variables to ensure that they do not overlap (i.e., that the various effects are separated in the terms of the summation equation). In particular, this difficulty arises from the complex nature of the forces of interaction between water and the solid matrix (including adsorption, exchangeable ion and capillary effects).

The difference in chemical potential between water in the soil and pure free water at the same temperature has been called the *moisture potential* of the soil (Taylor and Slatyer, 1960). The *total potential* given in our preceding section includes a gravitational term as well.

We shall now proceed to describe the various components of the total potential of soil water separately.

D. Gravitational Potential

Every body on the earth's surface is attracted toward the earth's center by a gravitational force equal to the weight of the body, that weight being the product of the body's mass by the gravitational acceleration. To raise a body against this attraction, work must be expended, and this work is stored by the raised body in the form of gravitational potential energy. The amount of this energy depends on the body's position in the gravitational force field.

The gravitational potential of soil water at each point is determined by the elevation of the point relative to some arbitrary reference level. For the sake of convenience, it is customary to set the reference level at the elevation of a pertinent point within the soil, or below the soil profile being considered, so that the gravitational potential can always be taken as positive or zero.

At a height z above a reference, the gravitational potential energy E_g of a mass M of water, occupying a volume V, is

$$E_g = Mgz = \rho_w Vgz \tag{3.9}$$

where ρ_w is the density of water and g the acceleration of gravity. Accordingly, the gravitational potential, in terms of the potential energy per unit mass, is

$$\phi_g = gz \tag{3.10}$$

and in terms of potential energy per unit volume

$$\phi_g = \rho_w gz \tag{3.11}$$

The gravitational potential is independent of the chemical and pressure conditions of soil water, but dependent only on relative elevation.

E. Pressure Potential

When soil water is at a hydrostatic pressure greater than atmospheric, its pressure potential is considered positive. When it is at a pressure lower than atmospheric (a subpressure commonly known as *tension* or *suction*, the pressure potential is considered negative. Thus, water under a free-water surface is at positive pressure potential, while water at such a surface is at zero pressure potential, and water which has risen in a capillary tube above that surface is characterized by a negative pressure potential. This principle is illustrated in Fig. 3.1.

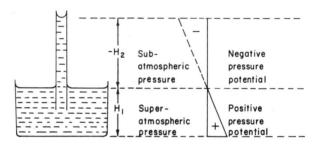

Fig. 3.1. Superatmospheric and subatmospheric pressures below and above a free-water surface.

The positive pressure potential which occurs below the ground water level has been termed the "submergence potential" (Rose, 1966). The hydrostatic pressure P of water with reference to the atmospheric pressure is

$$P = \rho g h \qquad (3.12)$$

where h is the submergence depth below the free-water surface (called the piezometric head by Terzaghi, 1951).

The potential energy[6] of this water is then

$$E = P \, dV \qquad (3.13)$$

and thus the submergence potential, taken as the potential energy per unit volume, is:

$$\phi_{ps} = P \qquad (3.14)$$

[6] For the special case of swelling media, Philip (1969) postulated the existence of an "overburden potential" in addition to the other manifestations of a pressure potential. Where any addition of water to the medium demands a local increase in bulk volume, the overburden potential is the work performed (per unit weight of added water) against gravity and the external load.

A negative pressure potential has often been termed "capillary potential," and more recently, "matric potential."[7] This potential of soil water results from the capillary and adsorptive forces due to the soil matrix. These forces attract and bind water in the soil and lower its potential energy below that of bulk water. As shown in our last chapter, capillarity results from the surface tension of water and its contact angle with the solid particles. In an unsaturated (three-phase) soil system, curved menisci form which obey the equation of capillarity

$$p_0 - p_c = P = \gamma \left(\frac{1}{R_1} + \frac{1}{R_2} \right) \qquad (3.15)$$

where p_0 is the atmospheric pressure, conventionally taken as zero; p_c the pressure of soil water, which can be smaller than atmospheric; P is the pressure deficit, or subpressure, of soil water; γ the surface tension of water; and R_1 and R_2 are the principal radii of curvature of a point on the meniscus.

If the soil were like a simple bundle of capillary tubes, the equations of capillarity might by themselves suffice to describe the relation of the negative pressure potential, or tension, to the radii of the soil pores in which the menisci are contained. However, in addition to the capillary phenomenon, the soil also exhibits adsorption, which forms hydration envelopes over the particle surfaces. These two mechanisms of soil–water interaction are illustrated in Fig. 3.2.

The presence of water in films as well as under concave menisci is most important in clayey soil and at high suctions, and it is influenced by the electric double layer and the exchangeable cations present. In sandy soils, adsorption is relatively unimportant and the capillary effect predominates. In general, however, the negative pressure potential results from the combined effect of the two mechanisms, which cannot easily be separated, since the capillary "wedges" are at a state of internal equilibrium with the adsorption "films," and the ones cannot be changed without affecting the others. Hence, the older term "capillary potential" is inadequate and the better term is "matric potential," as it denotes the total effect resulting from the affinity of the water to the whole matrix of the soil, including its pores and particle surfaces together.

---

[7] The terms "matric potential," "matric suction," and "soil-water suction" have been used interchangeably. According to the I.S.S.S. committee cited above, the matric suction is defined as "the negative gauge pressure, relative to the external gas pressure on the soil water, to which a solution identical in composition with the soil solution must be subjected in order to be in equilibrium through a porous membrane wall with the water in the soil." This definition implies the use of a *tensiometer*, which we shall describe in a subsequent section of this chapter.

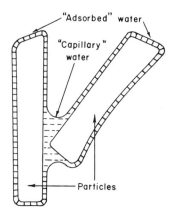

"Adsorbed" water

"Capillary" water

Particles

Fig. 3.2. Water in an unsaturated soil is subject to capillarity and adsorption, which combine to produce a matric suction.

Some soil physicists prefer to separate the positive pressure potential from the matric potential, assuming the two to be mutually exclusive. Accordingly, soil water may exhibit either of the two potentials, but not both simultaneously. Thus, unsaturated soil has no pressure potential, only a matric potential, which, however, is expressible in negative pressure units. This is really a matter of formality. There is an advantage in unifying the positive pressure potential and the matric potential (with the latter considered merely as a negative pressure potential) in that this unified concept allows one to consider the entire moisture profile in the field in terms of a single continuous potential extending from the saturated region into the unsaturated region, below and above the water table.

An additional factor which may affect the pressure of soil water is a possible change in the pressure of the ambient air. In general, this effect is negligible, as the atmospheric pressure remains nearly constant, small barometric pressure fluctuations notwithstanding (Peck, 1960). However, in the laboratory, the application of air pressure to change soil-water pressure or suction is a common practice. Hence, this effect was recognized and termed the *pneumatic potential* in a draft report by the International Committee on soil Physics Terminology (Aslyng *et al.*, 1962). In an unsaturated soil ϕ_p can be taken as equal to the sum of the matric (ϕ_m) and pneumatic (ϕ_a) potentials.

In the absence of solutes, the liquid and vapor phases in an unsaturated porous medium are related at equilibrium by

$$h = \exp(g\phi_m/RT) \tag{3.16}$$

where h is the relative humidity, R the gas constant for water vapor, and T the absolute temperature.

F. Osmotic Potential

The presence of solutes in soil water affects its thermodynamic properties and lowers its potential energy. In particular, solutes lower the vapor pressure of soil water. While this phenomenon may not affect mass liquid flow significantly, it does come into play whenever a membrane or diffusion barrier is present which transmits water more readily than salts. The osmotic effect is important in the interaction between plant roots and soil, as well as in processes involving vapor diffusion.

G. Quantitative Expression of Soil-Water Potential

The potential is expressible physically in at least three ways:

1. *Energy per unit mass*: This is often taken to be the fundamental expression of potential, using units of ergs per gram or joules per kilogram. The dimensions of energy per unit mass are L^2T^{-2}.

2. *Energy per unit volume*: Since water is a practically incompressible liquid, its density is almost independent of potential. Hence, there is a direct proportion between the expression of the potential as energy per unit mass and its expression as energy per unit volume. The latter expression yields the dimensions of pressure (for, just as energy can be expressed as the product of pressure by volume, so the ratio of energy to volume gives a pressure). This equivalent pressure can be measured in terms of dynes per square centimeter, bars, or atmospheres. The basic dimensions are those of force per unit area: $ML^{-1}T^{-2}$. This method of expression is convenient for the osmotic and pressure potentials, but is seldom used for the gravitational potential.

3. *Energy per unit weight (hydraulic head)*: Whatever can be expressed in units of hydrostatic pressure can also be expressed in terms of an equivalent hydraulic head, which is the height of a liquid column corresponding to the given pressure. For example, a pressure of 1 atm is equivalent to a vertical water column, or hydraulic head, of 1033 cm, and to a mercury head of 76 cm. This method of expression is certainly simpler, and often more convenient, than the previous methods. Hence, it is common to characterize the state of soil water in terms of the "total potential head," the "gravitational potential head," and the "pressure potential head," which are usually expressible in centimeters of water.

Accordingly, instead of

$$\phi = \phi_g + \phi_p \tag{3.17}$$

one could write

$$H = H_g + H_p \tag{3.18}$$

which reads: The total potential head of soil water (H) is the sum of the gravitational (H_g) and pressure (H_p) potential heads. H is also called the hydraulic head.

In attempting to express the negative pressure potential of soil water in terms of an equivalent hydraulic head, we must contend with the fact that this head may be as much as $-10,000$ or even $-100,000$ cm of water. To avoid the use of such cumbersomely large numbers, Schofield (1935) suggested the use of "pF" (by analogy with the pH acidity scale) which he defined as the logarithm of the negative pressure (tension, or suction) head in centimeters of water. A pF of 1 is, thus, a tension head of 10 cm H_2O, a pF of 3 is a tension head of 1000 cm H_2O; and so forth.

The use of various alternative methods for expressing the soil-water potential can be confusing to the uninitiated. It should be understood that these alternative expressions are in fact equivalent, and each method of

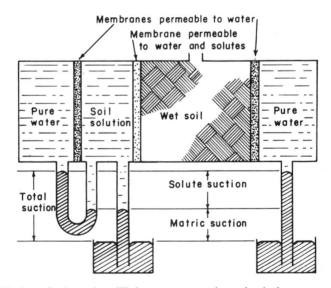

Fig. 3.3. In an isothermal equilibrium system, matric suction is the pressure difference across a membrane separating soil solution *in situ* from the same solution in bulk, the membrane being permeable to solution but not to solid particles or air; osmotic suction is the pressure difference across a semipermeable membrane separating bulk phases of free water and the soil solution; and total suction is the sum of the matric and osmotic suction values, and is thus the pressure difference across a semipermeable membrane separating pure water from a soil that contains solution. An ideal semipermeable membrane is permeable to water only. (After L. A. Richards, 1965).

expression can be translated directly into any of the other methods. If we use ϕ to designate the potential in terms of energy per mass, P for the potential in terms of pressure, and H for the potential head, then

$$\phi = \frac{P}{\rho_w} \tag{3.19}$$

$$H = \frac{P}{\rho_w g} = \frac{\phi}{g} \tag{3.20}$$

where ρ_w is the density of liquid water and g the acceleration of gravity.

A remark is in order concerning the use of the synonymous terms "tension" and "suction" in lieu of "negative (or) subatmospheric pressure." These are merely semantic devices to avoid the use of the unesthetic negative sign which generally characterizes the pressure of soil water, and to allow us to speak of the osmotic and matric potentials in positive terms. These two potentials, separately and in combination, are illustrated in Fig. 3.3.

H. Soil-Moisture Characteristic Curve

In a saturated soil at equilibrium with free water at the same elevation, the actual pressure is atmospheric, and hence the hydrostatic pressure and the suction (or tension) are zero.

If a slight suction, i.e., a water pressure slightly subatmospheric, is applied to water in a saturated soil, no outflow may occur until, as suction is increased, a certain critical value is exceeded at which the largest pore of entry begins to empty. This critical suction is called the *air-entry suction*. Its value is generally small in coarse-textured and in well-aggregated soils. However, since in coarse-textured soils the pores are often more nearly uniform in size, these soils may exhibit critical air-entry phenomena more distinctly and sharply than do fine-textured soils.

As suction is further increased, more water is drawn out of the soil and more of the relatively large pores, which cannot retain water against the suction applied, will empty out. Recalling the capillary equation[8] $(-P = 2\gamma/r)$, we can readily predict that a gradual increase in suction will result in the emptying of progressively smaller pores, until, at high suction values, only the very narrow pores retain water. Similarly, an increase in soil-water suction is associated with a decreasing thickness of the hydration envelopes covering the soil-particle surfaces. Increasing suction is thus associated with decreasing soil wetness. The amount of water remaining in the soil at

[8] The use of this equation assumes that the contact angle is zero and that soil pores are approximately cylindrical.

equilibrium is a function of the sizes and volumes of the water-filled pores and hence it is a function of the matric suction. This function is usually measured experimentally, and it is represented graphically by a curve known as the "soil-moisture-retention curve," or the "soil-moisture characteristic" (Childs, 1940).

Table 3.1

ENERGY LEVELS OF SOIL WATER

Soil-water potential				Soil-water suction[a]		Vapor pressure (Torr)	Relative humidity at 20°C (%)
per unit mass		per unit volume		Pressure (bars)	Head (cm H_2O)		
ergs/gm	joules/kg	(bars)	(cm H_2O)				
0	0	0	0	0	0	17.5350	100.00
-1×10^4	-1	-0.01	-10.2	0.01	10.2	17.5349	100.00
-5×10^4	-5	-0.05	-51.0	0.05	51.0	17.5344	99.997
-1×10^5	-10	-0.1	-102.0	0.1	102.0	17.5337	99.993
-2×10^5	-20	-0.02	-204.0	0.2	204.0	17.5324	99.985
-3×10^5	-30	-0.3	-306.0	0.3	306.0	17.5312	99.978
-4×10^5	-40	-0.4	-408.0	0.4	408.0	17.5299	99.971
-5×10^5	-50	-0.5	-510.0	0.5	510.0	17.5286	99.964
-6×10^5	-60	-0.6	-612.0	0.6	612.0	17.5273	99.965
-7×10^5	-70	-0.7	714.0	0.7	714.0	17.5260	99.949
-8×10^5	-80	-0.8	-816.0	0.8	816.0	17.5247	99.941
-9×10^5	-90	-0.9	918.0	0.9	918.0	17.5234	99.934
-1×10^6	-100	-1.0	-1020	1.0	1020	17.5222	99.927
-2×10^6	-200	-2	-1040	2	2040	17.5089	99.851
-3×10^6	-300	-3	-3060	3	3060	17.4961	99.778
-4×10^6	-400	-4	-4080	4	4080	17.4833	99.705
-5×10^6	-500	-5	-5100	5	5100	17.4704	99.637
-6×10^6	-600	-6	-6120	6	6120	17.4572	99.556

[a] In the absence of osmotic effects (soluble salts), soil-water suction equals matric suction; otherwise, it is the sum of matric and osmotic suctions.

As yet, no satisfactory theory exists for the prediction of the matric suction vs. wetness relationship from basic soil properties. The adsorption and pore-geometry effects are often too complex to be described by a simple model. Several empirical equations have been proposed which apparently describe the soil-moisture characteristic for some soils and within limited suction ranges. One such equation was advanced by Visser (1966):

$$\psi = a(f - \theta)^b/\theta^c \qquad (3.21)$$

where ψ is the matric suction, f is the porosity, θ is the volumetric wetness, and a, b, and c are constants. The actual use of this equation is hampered by the difficulty of evaluating its constants. Visser found that b varied between 0 and 10, a between 0 and 3, and f between 0.4 and 0.6.

Gardner et al. (1970) proposed the empirical relation

$$\psi = a\theta^{-b} \tag{3.22}$$

Constant b was found to have a value of 4.3 for a fine sandy loam. This relation fits only a limited range of the characteristic curve, but it may be useful in analyzing processes in which the water-content range is narrow (e.g., redistribution or internal drainage).

The amount of water retained at relatively low values of matric suction (say, between 0 and 1 bar of suction) depends primarily upon the capillary effect and the pore-size distribution, and hence is strongly affected by the structure of the soil. On the other hand, water retention in the higher suction range is due increasingly to adsorption and is thus influenced less by the structure and more by the texture and specific surface of the soil material. According to Gardner (1968), the water content at a suction of 15 bars (often taken to be the lower limit of soil moisture availability to plants) is fairly well correlated with the surface area of a soil and would represent, roughly, about 10 molecular layers of water if it were distributed uniformly over the particle surfaces.

It should be obvious from the foregoing that the soil-moisture characteristic curve is strongly affected by soil texture. The greater the clay content, in general, the greater the water content at any particular suction, and the more gradual the slope of the curve. In a sandy soil, most of the pores are relatively large, and once these large pores are emptied at a given suction, only a small amount of water remains. In a clayey soil, the pore-size distribution is more uniform, and more of the water is adsorbed, so that increasing the matric suction causes a more gradual decrease in water content (Fig. 3.4).

Soil structure also affects the shape of the soil-moisture characteristic curve, particularly in the low-suction range. The effect of compaction upon a soil is to decrease the total porosity, and, especially, to decrease the volume of the large interaggregate pores. This means that the saturation water content and the initial decrease of water content with the application of low suction are reduced. On the other hand, the volume of intermediate-size pores is likely to be somewhat greater in a compact soil (as some of the originally large pores have been squeezed into intermediate size by compaction), while the intraaggregate micropores remain unaffected and thus the curves for the compacted and uncompacted soil may be nearly identical

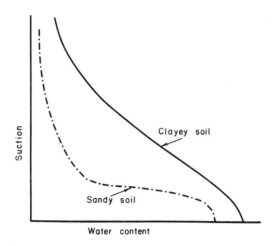

Fig. 3.4. The effect of texture on soil water retention.

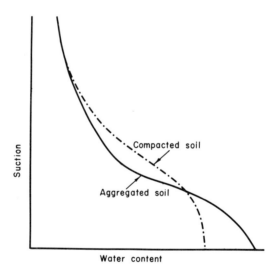

Fig. 3.5. The effect of structure on soil water retention.

at the high-suction range (Fig. 3.5). At the very-high-suction range, water is held primarily by adsorption, and hence retention is a textural rather than a structural attribute of the soil.

If two soil bodies differing in texture or in structure are equilibrated in contact with each other, they will after a time attain equal potential but generally exhibit a discontinuity in water content, corresponding to their different moisture-characteristic curves.

In a nonshrinking soil, the soil-moisture characteristic curve, once obtained, allows calculation of the effective pore-size distribution (i.e., the volumes of different classes of pore sizes). If an increase in matric suction from ψ_1 to ψ_2 results in the release of a certain volume of water, then that volume is evidently equal to the volume of pores having a range of effective radii between r_1 and r_2, where ψ_1 and r_1, and ψ_2 and r_2, are related by the equation of capillarity, namely $\psi = 2\gamma/r$.

An important fact which deserves to be stressed once again is that water in an unsaturated soil, being at subatmospheric pressure, will not tend to seep into the atmosphere. To flow spontaneously out of the soil and into the atmosphere, soil-water pressure must exceed atmospheric pressure. Similarly, for a fine-textured soil to drain into the large pores of an initially dry, coarse-textured layer, soil water must be at nearly atmospheric pressure.

The slope of the soil-moisture characteristic curve, which is the change of water content per unit change of matric potential, is generally termed the *differential* (or specific) *water capacity*[9] c:

$$c_\theta = \frac{d\theta}{d\phi_p}, \text{ or } c_\theta = -\frac{d\theta}{d\psi} \qquad (3.23)$$

This is an important property in relation to soil-moisture storage and availability to plants. The actual value of c depends upon the wetness range, the texture, and the hysteresis effect.

I. Hysteresis

The relation between matric potential and soil wetness is not generally a unique and single-valued one. This relation can be obtained in two ways: (1) in *desorption*, by taking an initially-saturated sample and applying increasing suction to gradually dry the soil while taking successive measurements of water content versus suction; and (2) in *sorption*, by gradually wetting up an initially dry soil sample while reducing the suction. Each of these two methods yields a continuous curve, but the two curves will in general not be identical. The equilibrium soil wetness at a given suction is greater in desorption (drying) than in sorption (wetting). This dependence of the equilibrium

---

[9] This by analogy with the differential heat capacity, which is the change in heat content per unit change in the thermal potential (temperature). However, while the latter is fairly constant with temperature for many materials, the differential water capacity in soils is strongly dependent on the matric potential.

content and state of soil water upon the direction of the process leading up to it is called hysteresis.[10]

Basic studies of soil-water hysteresis phenomena were reported by Haines (1930), Miller and Miller (1955 a,b; 1956), Poulovassilis (1962), Philip (1964), Topp and Miller (1966), Bomba (1968), and Topp (1969).

Figure 3.6 shows a typical soil-moisture characteristic curve and illustrates the hysteresis effect in the soil-water equilibrium relationship.

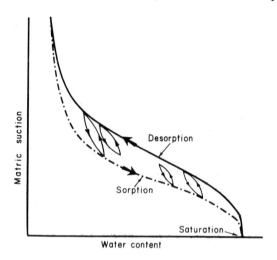

Fig. 3.6. The suction–water-content curves in sorption and desorption. The intermediate loops are "scanning curves," indicating transitions between the main branches.

The hysteresis effect may be attributed to several causes: (1) the geometric nonuniformity of the individual pores (which are generally irregularly shaped voids interconnected by smaller passages), which results in the "inkbottle" effect, illustrated in Fig. 3.7. (2) The contact-angle effect, mentioned in Chapter 2, by which the contact angle is greater and hence, the radius of curvature is greater, in an advancing meniscus than in the case of a receding one.[11] A given water content will tend therefore to exhibit greater suction

[10] A description of hysteresis in more general physical terms was offered by Poulovassilis (1962): If a physical property Y depends upon an independent variable X, then it may occur that the relationship between Y and X is unique, and, in particular, independent of whether X is increasing or decreasing. Such a relationship is reversible. Many physical properties are, however, irreversible, and even when the changes of X are made very slowly the curve for increasing X does not coincide with that for decreasing X. This phenomenon, called hysteresis, is commonly observed in magnetism, for instance.

[11] Contact-angle hysteresis can arise because of surface roughness, the presence and distribution of adsorbed impurities on the solid surface, and the mechanism by which liquid molecules adsorb or desorb when the interface is displaced.

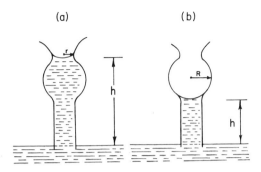

Fig. 3.7. "Inkbottle" effect determines equilibrium height of water in a variable-width pore: (a) in capillary drainage (desorption); (b) in capillary rise (sorption).

in desorption than in sorption. (3) Entrapped air, which further decreases the water content of newly wetted soil. Failure to attain true equilibrium can accentuate the hysteresis effect. (4) Swelling, shrinking, or aging phenomena, which result in differential changes of soil structure, depending on the wetting and drying history of the sample (Hillel and Mottes, 1964). The gradual solution of air, or the release of dissolved air from soil water, can also have a differential effect upon the suction–wetness relationship in wetting and drying systems.

Of particular interest is the inkbottle effect. Consider the hypothetical pore shown in Fig. 3.7. This pore consists of a relatively wide void of radius R, bounded by narrow channels of radius r. If initially saturated, this pore will drain abruptly the moment the suction exceeds ψ_r, where $\psi_r = 2\gamma/r$. For this pore to rewet, however, the suction must decrease to below ψ_R, where $\psi_R = 2\gamma/R$, whereupon the pore abruptly fills. Since $R > r$, it follows that $\psi_r > \psi_R$. Desorption depends on the narrow radii of the connecting channels, whereas sorption depends on the maximum diameter of the large pores. These discontinuous spurts of water, called "Haines jumps," can be observed readily in coarse sands. The hysteresis effect is in general more pronounced in coarse-textured soils in the low-suction range, where pores may empty at a much larger suction than that at which they fill.

In the past, hysteresis was generally disregarded in the theory and practice of soil physics. This may be justifiable in the case of processes entailing monotonic wetting (e.g., infiltration) or drying (e.g., evaporation). But the hysteresis effect may be important in the case of composite processes in which wetting and drying occur simultaneously or sequentially in various parts of the soil profile (e.g., redistribution). It is possible to have two soil layers of identical texture and structure but different wetness at equilibrium with each other or at identical energy states if their wetting histories have been different. Furthermore, hysteresis can affect the dynamic, as well as the static, properties of the soil (Poulovassilis, 1969).

The two complete characteristic curves, from saturation[12] to dryness and vice versa, are called the *main branches* of hysteresis. When a partially wetted soil commences to drain, or when a partially desorbed soil is rewetted, the relation of suction to moisture content follows some intermediate curve as it moves from one main branch to the other. Such intermediate spurs are called *scanning curves*. Cyclic changes often entail wetting and drying scanning curves, which may form loops between the main branches (Fig. 3.6). The ψ vs. θ relationship can thus become very complicated.

J. Measurement of the Soil Water Content (Wetness)

The need to determine the amount of water contained in the soil arises frequently in many soil and hydrological investigations. This information is requisite for understanding the soil's chemical, mechanical, and hydrological behavior, and the growth response of plants. There are direct and indirect methods to measure soil moisture, and several alternative ways to express it quantitatively. There is, therefore, no universally recognized standard method of measurement and no uniform way to compute and present the results. A review of the most common methods was given by Gardner (1965).

Soil wetness is usually expressed as a dimensionless ratio of water mass to dry soil mass, or of water volume to total soil volume. These ratios are usually multiplied by 100 and reported as percentages by mass or by volume. We shall proceed to describe, briefly, some of the most prevalent methods for this determination.

1. SAMPLING AND DRYING

The traditional ("gravimetric") method of measuring the water content by mass consists of removing a sample (e.g., by augering) and of determining its "moist" and "dry" weights (the latter after drying the sample to constant weight in an oven at 105°C). The gravimetric wetness (or mass wetness) is the ratio of the weight loss in drying to the dry weight of the sample (mass and weight being proportional). Occasionally, however, the mass ratio of the water to the wet soil is used. The mass (or weight) wetness on the dry-soil basis (w_{md}) and the mass wetness on the wet-soil basis (w_{mw}) can be converted to each other by the following equations:

---

[12] In actual practice, it is very difficult to obtain complete saturation in sorption by reducing the suction to zero when the ambient air is at atmospheric pressure, owing to the entrapment of air in soil pores.

$$w_{md} = \frac{w_{mw}}{1 - w_{mw}} \qquad (3.24)$$

$$w_{mw} = \frac{w_{md}}{1 + w_{md}} \qquad (3.25)$$

To obtain the volumetric wetness θ from the gravimetric determination, a separate measurement must be made of the bulk density ρ_b, and the following equation is used:

$$\theta = \left(\frac{\rho_b}{\rho_w}\right) w_{md} \qquad (3.26)$$

The measurement of bulk density, particularly in the field, is difficult and subject to errors. The gravimetric method itself, depending as it does on sampling, transporting, and repeated weighings, entails inherent errors. It is

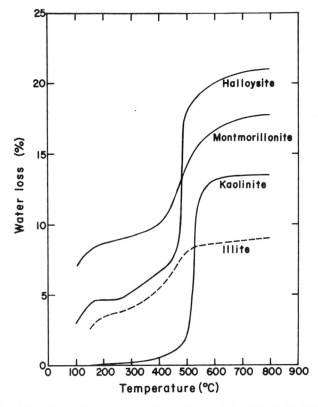

Fig. 3.8. Dehydration curves of clay minerals (after Marshall, 1964).

also laborious and time-consuming, since a period of at least 24 hr is usually allowed for complete drying. The standard method of oven drying is also arbitrary. Some clays may still contain appreciable amounts of adsorbed water even at 105°C (Fig. 3.8). On the other hand, some organic matter may oxidize and decompose at this temperature so that the weight loss may not be due entirely to the evaporation of water.

The errors of the gravimetric method can be reduced by increasing the sizes and number of samples. However, the sampling method is destructive and may disturb an observation or experimental plot sufficiently to distort the results. For these reasons, many workers prefer indirect methods, which permit making frequent or continuous measurements at the same points, and, once the equipment is installed and calibrated, with much less time and labor.

2. ELECTRICAL RESISTANCE

The electrical resistance of a soil volume depends not only upon its water content, but also upon its composition, texture, and soluble-salt concentration. On the other hand, the electrical resistance of porous bodies placed in the soil and left to equilibrate with soil moisture can sometimes be calibrated against the water content. Such units (generally called electrical resistance blocks) generally contain a pair of electrodes embedded in gypsum (Bouyoucos and Mick, 1940), nylon, or fiberglas (Colman and Hendrix, 1949).

Porous blocks embedded in the soil tend to equilibrate with the soil moisture (matric) suction, rather than with the soil-moisture content directly. Different soils can have greatly differing wetness vs. suction relationships (e.g., a sandy soil may retain less than 5% moisture at, say, 15-bar suctions, whereas a clayey soil may retain three or four times as much). Hence, calibration of porous blocks against suction (tension) is basically preferable to calibration against soil wetness, particularly when the soil used for calibration is a disturbed sample differing in structure from the soil *in situ.*

The equilibrium of porous blocks with soil moisture may be affected by hysteresis, i.e., the direction of change. Furthermore, the hydraulic properties of the blocks (or inadequate contact with the soil) may impede the rapid attainment of equilibrium and cause a time lag between the state of water in the soil and the state of water being measured in the block. This effect, as well as the sensitivity, may not be constant over the entire range of variation in soil wetness which we may desire to measure.

The electrical conductivity of most porous blocks is due primarily to the permeating fluid rather than to the solid matrix. Thus, it depends upon the electrolytic solutes present in the fluid as well as upon the volume content of the fluid. Blocks made of such inert materials as fiberglas, for instance, are highly sensitive to even small variations in salinity of the soil solution. On the other hand, blocks made of plaster of Paris (gypsum) maintain a nearly constant electrolyte concentration corresponding primarily to that of a saturated solution of calcium sulfate. This tends to mask, or buffer, the effect of small or even moderate variations in the soil solution (such as those due to fertilization or low levels of salinity). However, since gypsum is soluble, these blocks eventually deteriorate in the soil.

For these and other reasons (e.g., temperature sensitivity), the evaluation of soil wetness by means of electrical resistance blocks is likely to be of limited accuracy. Soil moisture blocks have been found to be more dependable in the drier than in the wetter range (Johnson, 1962). An advantage of these blocks is that they can be connected to a recorder to obtain a continuous record of soil-moisture changes *in situ*.

3. NEUTRON SCATTERING

In recent years, this method has gained widespread acceptance for monitoring profile water content in the field. Its advantage is that it allows rapid and periodically repeated measurements in the same locations and depths of the volumetric wetness of a representative volume of soil.

The instrument known as a neutron moisture meter consists of two principal parts: (a) a *probe*, which is lowered into an access tube inserted vertically into the soil, and which contains a source of fast neutrons and a detector of slow neutrons; (b) a *scaler* or *ratemeter* (usually battery-powered and portable) to monitor the flux of slow neutrons, which is proportional to the soil water content. The fast-neutron source may be a 2–5 millicurie mixture of radium and beryllium (which also emits hazardous γ-radiation), or a mixture of americium and beryllium (with less-hazardous γ-radiation). The source materials are chosen for their longevity (e.g., radium–beryllium has a half-life of 1620 years) so that they can be used for a number of years without an appreciable change in radiation flux.

The fast neutrons are emitted radially into the soil, where they encounter and collide elastically with various atomic nuclei, and gradually lose some of their kinetic energy. The average loss of energy is maximal when a neutron collides with a particle of a mass nearly equal to its own. Such particles are the hydrogen nuclei of water. The average number of collisions required to slow a neutron from 2 MeV to thermal energies is 18 for hydrogen, 114 for carbon, 150 for oxygen, and $9N + 6$ for nuclei of larger mass number N (Weinberg and Wigner, 1958). In practice, it has been found that the attenuation of fast neutrons in the soil is proportional to the hydrogen content of the soil. The slowed ("thermal") neutrons scatter randomly in the soil, forming a cloud around the probe. Some of these return to the probe, where they are counted by a detector of slow neutrons. The detector cell is usually filled with BF_3 gas. When a thermalized neutron encounters a ^{10}B nucleus and is absorbed, an alpha particle (the helium nucleus) is emitted, creating an electrical pulse on a charged wire. The number of pulses over a measured time interval is counted by a scaler, or indicated by a ratemeter (Gardner and Kirkham, 1952; van Bavel *et al.*, 1956).

The effective volume of soil in which the water content is measured depends upon the concentration of the hydrogen nuclei, i.e., upon the per-volume wetness of the soil, as well as upon the energy of the emitted fast neutrons. With the commonly used radium–beryllium sources, the soil volume measured is in effect a sphere, which in a wet soil is perhaps 15 cm in diameter, but in a relatively dry soil may be as great as 50 cm or more (de Vries and King, 1961; van Bavel *et al.*, 1961). This low degree of spatial resolution makes the instrument unsuitable for detection of water-content discontinuities (e.g., wetting fronts or boundaries between layers), or for measurements close to the soil surface. The relatively large volume monitored can, however, be an advantage in water-balance studies, for instance, as such a volume is generally more representative of the field soil than a small sample.

Methods of calibrating the neutron moisture meter were described by Holmes (1956) and Holmes and Jenkinson (1959). In most soils, it is possible to obtain a nearly linear dependence of the count rate upon the volumetric wetness of the soil.

Improper use of the equipment can be hazardous. The danger from exposure to radiation depends upon the strength of the source, the distance from the source to the operator, and the duration of exposure. A protective shield (generally consisting of lead and paraffin or polyethylene in a cylindrically shaped unit with a central hole to accommodate the probe) is an essential component of the equipment, and it also serves as a standard absorber for checking the readings. With reasonable care and attention to safety rules, the equipment can be used safely.

4. OTHER METHODS

Additional techniques for the measurement of soil wetness include gamma-ray attenuation (Gurr, 1962), the dependence of soil thermal properties on water content, and the use of ultrasonic waves. Few of these methods are as yet universally applicable or practical for routine use in the field.

K. Measurement of Soil Moisture Potential

The measurement of soil water content, or wetness, though essential in many soil physical and engineering investigations, is not sufficient to provide a description of the state of soil water. To obtain such a description, the

evaluation of the energy status of soil water (soil moisture potential, or suction)
δ is necessary. In general, the two properties, wetness and potential, should
each be measured directly, as the translation of one to the other on the
basis of calibration curves of soil samples has been found unreliable
(Stolzy et al., 1959).

Total soil moisture potential is often thought of as the sum of matric and
osmotic (solute) potentials, and is a useful index for characterizing the energy
status of soil water with respect to plant water uptake. The sum of the matric
and gravitational (elevation) heads is generally called the hydraulic head
(or hydraulic potential), and is useful in evaluating the direction and intensity
of the water-moving forces in the soil profile. Methods are available for
measuring matric potential as well as total soil potential, separately or together.
A schematic representation of these suction terms is shown in Fig. 3.3. To
measure matric potential in the field, an instrument known as the tensiometer is used. To measure soil moisture potential, freezing-point-depression
measurements (Richards and Campbell, 1949), and vapor-pressure measurements of soil water by means of thermocouple psychrometers (Richards and
Ogata, 1958; Rawlins, 1966; Rawlins and Dalton, 1967; Dalton and Rawlins,
1968) have been used.

At equilibrum, the potential of soil water is equal to the potential of the
water vapor in the ambient atmosphere. If thermal equilibrium is assured,
and the gravitional effect is neglected, the vapor potential is equal to the sum
of the matric and osmotic potentials, since air acts as an ideal semi-permeable
membrane in allowing only water molecules to pass (provided the solutes
are non volatile). At room temperature, relative humidity of the air is related
to the potential by (Bolt and Frissel, 1960):

$$pF = 6.5 + \log (2 - \log R.H.) \tag{3.27}$$

where $pF = \log$ (osmotic potential + matric potential) when these potentials
are expressed as cm of water column; and $R.H. =$ relative humidity.

We shall now describe the tensiometer, which has won widespread acceptance as a practical device for the *in situ* measurement of matric suction,
hydraulic head, and hydraulic gradients.

1. THE TENSIOMETER

The essential parts of a tensiometer are shown in Fig. 3.9. The tensiometer
consists of a porous cup, generally of ceramic material, connected through a
tube to a manometer, with all parts filled with water. When the cup is placed
in the soil where the suction measurement is to be made, the bulk water
inside the cup comes into hydraulic contact and tends to equilibrate with
soil water through the pores in the ceramic walls. When initially placed in
the soil, the water contained in the tensiometer is generally at atmospheric

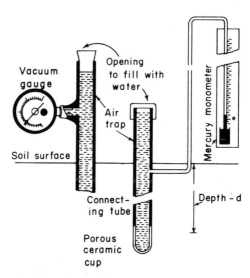

Fig. 3.9. Schematic illustration of the essential parts of a tensiometer (After S. J. Richards, 1965).

pressure. Soil water, being generally at subatmospheric pressure, exercises a suction which draws out a certain amount of water from the rigid and airtight tensiometer, thus causing a drop in its hydrostatic pressure. This pressure is indicated by a manometer, which may be a simple water- or mercury-filled U-tube, a vacuum gauge, or an electrical transducer.

A tensiometer left in the soil for a long period of time tends to follow the changes in the matric suction of soil water. As soil moisture is depleted by drainage or plant uptake, or as it is replenished by rainfall or irrigation, corresponding readings on the tensiometer gauge occur. Owing to the hydraulic resistance of the cup and the surrounding soil, or of the contact zone between the cup and the soil, the tensiometer response may lag behind suction changes in the soil. This lag time can be minimized by the use of a null-type device (Miller, 1951; Leonard and Low, 1962) or of a transducer type manometer with rigid tubing, so that practically no flow of water takes place as the tensiometer adjusts to changes in the soil matric suction.

Since the porous cup walls of the tensiometer are permeable to both water and solutes, the water inside the tensiometer assumes the same solute composition and concentration as soil water, and the instrument does not indicate the osmotic suction of soil water (unless equipped with some type of an auxiliary salt sensor).

Suction measurements by tensiometry are generally limited to matric suction values of below 1 atm. This is due to the fact that the vacuum gauge or manometer measures a partial vacuum relative to the external atmospheric pressure, as well as to the general failure of water columns in macroscopic systems to withstand tensions exceeding 1 atm. Furthermore, as the ceramic material is generally made of the most permeable and porous material possible, too high a suction may cause air entry into the cup, which would equalize the internal pressure to the atmospheric. Under such conditions, soil suction might continue to increase even while the tensiometer fails to show it.

In practice, the useful limit of most tensiometers is at about 0.8 bar of maximal suction. To measure higher suctions, the use of an osmometer with a semipermeable membrane at the wall has been proposed (Peck and Rabbidge, 1969), but the practical application of this instrument is still in the experimental stage. The limited range of suction measurable by the tensiometer is not as serious as it may seem at first sight. Though the suction range of 0–0.8 bar is but a small part of the total range of suction variation encountered in the field, it generally encompasses the greater part of the soil wetness range. Richards and Marsh (1961) have shown that in many agricultural soils the tensiometer range accounts for more than 50% (and in coarse-textured soils 75% or more) of the amount of soil water taken up by plants. Thus, where soil management (particularly in irrigation) is aimed at maintaining low-suction conditions which are most favorable for plant growth, tensiometers are definitely applicable.

Despite their fundamental and practical shortcomings, tensiometers are practical instruments, available commercially, and, when operated and maintained by a skilled worker, are capable of providing reliable data on the *in situ* state of soil-moisture profiles and their changes with time.

Tensiometers have been found useful in guiding the timing of irrigation of field and orchard crops, as well as of potted plants (Richards and Marsh, 1961). A general practice is to place a tensiometer at one or more soil depths representing the root zone, and to irrigate when the tensiometer indicates that the matric suction has reached some prescribed value. The use of several tensiometers at different depths can indicate the amount of water needed in irrigation, and can also allow calculation of the hydraulic gradients in the soil profile (L. A. Richards, 1955). If $\psi_1, \psi_2, \psi_3 \ldots, \psi_n$ are the matric suction values in centimeters of water head ($\cong$ millibars) at depths $d_1, d_2, d_3 \ldots, d_n$ measured in centimeters below the surface, the average hydraulic gradient i between depths d_n and d_{n+1} is

$$i = [(\psi_{n+1} + d_{n+1}) - (\psi_n + d_n)]/(d_{n+1} - d_n) \qquad (3.28)$$

Measurement of the hydraulic gradient is particularly important in the region below the root zone, where the direction and magnitude of water movement cannot easily be ascertained otherwise.

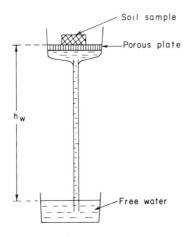

Fig. 3.10. Tension plate assembly for equilibrating a soil sample with a known matric suction value. This assembly is applicable in the range of 0–1 bar only.

2. MEASUREMENT OF THE SOIL-MOISTURE CHARACTERISTIC CURVES

The functional relation between soil wetness and matric suction is often determined by means of a tension plate assembly (Fig. 3.10) in the low suction (< 1 bar) range, and by means of a pressure plate or pressure membrane apparatus (Fig. 3.11) in the higher suction range. These instruments allow the application of successive suction values and the repeated measurement of the equilibrium soil wetness at each suction.

The maximum suction value obtainable by porous-plate devices is limited to 1 bar if the soil air is kept at atmospheric pressure and the pressure difference across the plate is controlled either by vacuum or by a hanging water column. Matric suction values considerably greater than 1 bar (say, 20 bars or even more) can be obtained by increasing the pressure of the air phase. This requires placing the porous-plate assembly inside a pressure chamber, as shown in Fig. 3.11. The limit of matric suction obtainable with

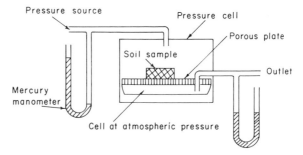

Fig. 3.11. Pressure plate apparatus for moisture-characteristic measurements in the high-suction range. The lower side of the porous plate is in contact with water at atmospheric pressure. Air pressure is used to extract water from initially saturated soil samples.

such a device is determined by the design of the chamber (i.e., its safe working pressure) and by the maximal air-pressure difference the saturated porous plate can bear without allowing air to bubble through its pores. Ceramic plates generally do not hold pressures greater than about 20 bars, but cellulose acetate membranes can be used with pressures exceeding 100 bars.

Soil moisture retention in the low-suction range (0–1 bar) is strongly influenced by soil structure and pore-size distribution. Hence, measurements made with disturbed samples (e.g., dried, screened, and artificially packed samples) cannot be expected to represent field conditions. The use of un-disturbed soil cores is therefore preferable. On the other hand, soil moisture retention in the high-suction range is due primarily to adsorption and is thus correlated with the specific surface of the soil material rather than with its structure.

As mentioned earlier, the soil-moisture characteristic is hysteretic. Ordinarily, the desorption curve is measured by gradually and monotonically decreasing the water content of initially saturated samples. The resulting curve, often called the soil-moisture-release curve, is applicable to processes involving drainage, evaporation, or plant extraction of soil moisture. On the other hand, the sorption curve is needed whenever infiltration or wetting processes are studied. Modified apparatus is required for the measurement of wetness vs. suction during sorption (Tanner and Elrick, 1958).

L. Summary

Two types of information are generally needed in the study of soil-water phenomena: the quantity of water contained in the soil and the energy status of soil water (i.e., soil wetness and soil moisture suction). While these can be measured independently, they are functionally related. This relationship, which is affected by hysteresis, is described by the "soil-moisture character-istic curve," variously termed the "retention," "release," "sorption," or "desorption" curve. By whatever name, it is of fundamental importance in soil physics, as it expresses the influence of structure, porosity, pore-size distribution, and adsorption on the state of soil water. This state and how it varies in the profile, in turn, determine the direction and influence the rate of soil-moisture movement and uptake by plants.

4 *Flow of Water in Saturated Soil*

A. Laminar Flow in Narrow Tubes

Before we enter into a discussion of flow in so complex a medium as soil, it might be helpful to consider some basic physical phenomena associated with fluid flow in narrow tubes.

Early theories of fluid dynamics were based on the hypothetical concepts of a "perfect" fluid, i.e., one that is both frictionless and incompressible. In the flow of a perfect fluid, contacting layers can exhibit no tangential forces (shearing stresses), only normal forces (pressures). Such fluids do not in fact exist. In the flow of real fluids, adjacent layers do transmit tangential stresses (drag), and the existence of intermolecular attractions causes the fluid molecules in contact with a solid wall to adhere to it rather than slip over it. The flow of a real fluid is associated with the property of viscosity, defined in Chapter 2.

We can visualize the nature of viscosity by considering the motion of a fluid between two parallel plates, one at rest, the other moving at a constant velocity (Fig. 4.1). Experience shows that the fluid adheres to both walls, so that its velocity at the lower plate is zero, and that at the upper plate is equal to the velocity of the plate. Furthermore, the velocity distribution in the fluid between the plates is linear, so that the fluid velocity is proportional to the distance y from the lower plate.

To maintain the relative motion of the plates at a constant velocity, it is necessary to apply a tangential force, that force having to overcome the frictional resistance in the fluid. This resistance, per unit area of the plate, is proportional to the velocity of the upper plate U and inversely proportional

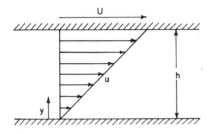

Fig. 4.1. Velocity distribution in a viscous fluid between two parallel flat plates, with the upper plate moving at a velocity U relative to the lower plate.

to the distance h. The shearing stress at any point τ_s, is proportional to the velocity gradient du/dy. The viscosity v is the proportionality factor between τ_s and du/dy[1]:

$$\tau_s = v \frac{du}{dy} \tag{4.1}$$

We can now apply these relationships to describe flow through a straight, cylindrical tube with a constant diameter $D = 2R$ (Fig. 4.2). The velocity is zero at the wall (because of adhesion), maximal on the axis, and constant on cylindrical surfaces which are concentric about the axis. Adjacent cylindrical *laminae*, moving at different velocities, slide over each other. A parallel

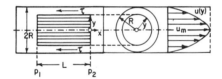

Fig. 4.2. Laminar flow through a cylindrical tube.

motion of this kind is called *laminar*. Fluid movement in a horizontal tube is generally caused by a pressure gradient acting in the axial direction. A fluid "particle," therefore, is accelerated by the pressure gradient and retarded by the frictional resistance.

Now let us consider a coaxial fluid cylinder of length L and radius y. For flow velocity to be constant, the pressure force acting on the face of the cylinder $\Delta p \pi y^2$, where $\Delta p = p_1 - p_2$, must be equal to the frictional resistance due to the shear force $2\pi y L \tau_s$ acting on the circumferential area. Thus,

$$\tau_s = \frac{\Delta p}{L} \frac{y}{2}$$

[1] Equation (4.1) bears an analogy to Hooke's law of elasticity. In an elastic solid, the shearing stress is proportional to the strain, whereas in a viscous fluid, the shearing stress is proportional to the time-rate of the strain.

Recalling Eq. (4.1),

$$\tau_s = - \, v \, \frac{du}{dy}$$

(the negative sign arises because in this case u decreases with y), we obtain

$$\frac{du}{dy} = - \frac{\Delta p}{vL} \frac{y}{2}$$

which, upon integration, gives

$$u(y) = \frac{\Delta p}{vL} \left(c - \frac{y^2}{4} \right)$$

The constant of integration c is evaluated from the boundary condition of no slip at the wall, that is, $u = 0$ at $y = R$, so that $c = R^2/4$.

Therefore,

$$u(y) = \frac{\Delta p}{4vL} (R^2 - y^2) \tag{4.2}$$

Equation (4.2) indicates that the velocity is distributed parabolically over the radius, with the maximum velocity u_{max} being on the axis ($y = 0$):

$$u_{max} = \frac{\Delta p R^2}{4vL}$$

The discharge Q, being the volume flowing through a section of length L per unit time, can now be evaluated. The volume of a paraboloid of revolution is $\frac{1}{2}$ (base × height), hence

$$Q = \frac{\pi}{2} R^2 u_{max} = \frac{\pi R^4 \Delta p}{8vL} \tag{4.3}$$

This equation, known as Poiseuille's law, indicates that the volume flow rate is proportional to the pressure drop per unit distance ($\Delta p/L$) and the fourth power of the radius of the tube.

The mean velocity over the cross section is

$$\bar{u} = \frac{\Delta p R^2}{8vL} = \left(\frac{R^2}{av} \right) \nabla p \tag{4.4}$$

where ∇p is the pressure gradient. Parameter a, equal to 8 in a circular tube, varies with the shape of the conducting passage.

Laminar flow prevails only at relatively low flow velocities and in narrow tubes. As the radius of the tube and the flow velocity are increased, the point is reached at which the mean flow velocity is no longer proportional to the pressure drop, and the parallel laminar flow changes into a turbulent flow with fluctuating eddies. Conveniently, however, laminar flow is the rule rather than the exception in most water-flow processes taking place in soils, because of the narrowness of soil pores (see discussion of Reynolds number, Section H).

B. Darcy's Law

Were the soil merely a bundle of straight and smooth tubes, each uniform in radius, we could assume the overall flow rate to be equal to the sum of the separate flow rates through the individual tubes. Knowledge of the size distribution of the tube radii could then enable us to calculate the total flow through a bundle caused by a known pressure difference, using Poiseuille's equation.

Unfortunately from the standpoint of physical simplicity, however, soil pores do not resemble uniform, smooth tubes, but are highly irregular, tortuous, and interconnected. Flow through soil pores is limited by numerous constrictions, or "necks," and occasional "dead-end" spaces. Hence, the actual geometry and flow pattern of a typical soil specimen is too complicated to be described in microscopic detail, as the fluid velocity varies drastically from point to point, even along the same passage. For this reason, flow through complex porous media is generally described in terms of a macroscopic flow velocity vector, which is the overall average of the microscopic velocities over the total volume of the soil. The detailed flow pattern is thus ignored, and the conducting body is treated as though it were a uniform medium, with the flow spread out over the entire cross section, solid and pore space alike.[2]

Let us now examine the flow of water in a macroscopically uniform, saturated soil body, and attempt to describe the quantitative relations connecting the rate of flow, the dimensions of the body, and the hydraulic conditions at the inflow and outflow boundaries.

Figure 4.3 shows a horizontal column of soil, through which a steady flow of water is occurring from left to right, from an upper reservoir to a lower one, in each of which the water level is maintained constant.

Experience shows that the discharge rate Q, being the volume V flowing through the column per unit time t, is directly proportional to the

---

[2] An implicit assumption here is that the soil volume taken is sufficiently large relative to the pore sizes and microscopic heterogeneities to permit the averaging of velocity and potential over the cross-section.

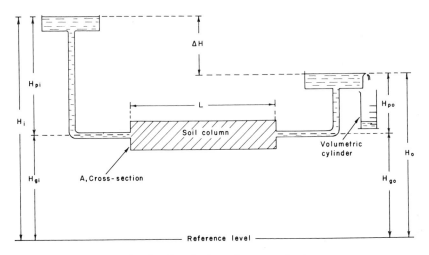

Fig. 4.3. Flow in a horizontal, saturated column.

cross sectional area and to the hydraulic-head drop ΔH, and inversely proportional to the length of the column L:

$$Q = \frac{V}{t} \propto \frac{A\,\Delta H}{L}$$

The usual way to determine the hydraulic head drop across the system is to measure the head at the inflow boundary (H_i) and at the outflow boundary (H_o), relative to some reference level. ΔH is the difference between these two heads:

$$\Delta H = H_i - H_o$$

Obviously, no flow occurs in the absence of a hydraulic head difference, i.e., when $\Delta H = 0$.

The head drop per unit distance in the direction of flow ($\Delta H / L$) is the *hydraulic gradient*, which is, in fact, the driving force. The specific discharge rate Q/A (i.e., the volume of water flowing through a cross-sectional area A per time t) is called the *flux density* (or simply the *flux*), and indicated by q. Thus, the flux is proportional to the hydraulic gradient:

$$q = \frac{Q}{A} = \frac{V}{At} \propto \frac{\Delta H}{L}$$

The proportionality factor K is generally designated as the *hydraulic conductivity*:

$$q = K\frac{\Delta H}{L} \tag{4.5}$$

This equation is known as Darcy's law, after Henri Darcy, the French engineer who first obtained it over a century ago in the course of his classic investigation of seepage rates through sand filters in the city of Dijon (Darcy, 1856; Hubbert, 1956).

Where flow is unsteady (i.e., the flux changing with time) or the soil non-uniform, the hydraulic head may not decrease linearly along the direction of flow. Where the hydraulic head gradient or the conductivity is variable, we must consider the localized gradient, flux, and conductivity values rather than overall values for the soil system as a whole. A more exact and generalized expression of the Darcy law is, therefore, in differential form. Slichter (1899) generalized Darcy's law for saturated porous media into a three-dimensional macroscopic differential equation of the form.[3]

$$\mathbf{q} = -K \nabla H \tag{4.6}$$

Stated verbally, this law indicates that the flow of a liquid through a porous medium is in the direction of, and at a rate proportional to, the driving force acting on the liquid (being the hydraulic gradient) and also proportional to the property of the conducting medium to transmit the liquid (namely, the conductivity).[4]

[3] $K \nabla H$ is the product of a scalar (K) and a vector (∇H), hence the flux $\mathbf{q}$ is a vector, the direction of which is determined by ∇H. This direction in an isotropic medium is orthogonal to surfaces of equal hydraulic potential, H.

[4] Ultimately, both Poiseuille's and Darcy's laws rest upon the more general *Navier–Stokes law*, which describes the flow of viscous fluids and forms the basis of the science of fluid mechanics. For an incompressible fluid in isothermal flow,

$$\frac{\partial \mathbf{u}}{\partial t} + (\mathbf{u} \cdot \nabla)\mathbf{u} = -\nabla \phi + \nu_k \nabla^2 \mathbf{u}$$

where $\mathbf{u}$ is the vector flow velocity, t time, ν_k the kinematic viscosity, and ϕ the potential including both the pressure term and the potential due to external or body forces (e.g., gravity).

This can be written as a set of three simultaneous differential equations of the type

$$\rho \left(\frac{\partial \mathbf{u}}{\partial t} + \mathbf{u} \frac{\partial \mathbf{u}}{\partial x} + \mathbf{v} \frac{\partial \mathbf{u}}{\partial y} + \mathbf{w} \frac{\partial \mathbf{u}}{\partial z} \right) = \nu_k \left(\frac{\partial^2 \mathbf{u}}{\partial x^2} + \frac{\partial^2 \mathbf{u}}{\partial y^2} + \frac{\partial^2 \mathbf{u}}{\partial z^2} \right) - \frac{\partial p}{dx} + X$$

(inertial terms)   (viscosity terms)   (pressure   (force
                                                                          gradient)   term)

where $\mathbf{u}$, $\mathbf{v}$, and $\mathbf{w}$ are component velocities along axes x, y, and z, respectively. Here, ρ is density, p pressure, and X represents the x component of external forces such as gravity. Similar relations exist for $\mathbf{v}$ and $\mathbf{w}$.

An analysis of how Poiseuille's and Darcy's laws relate to the Navier–Stokes law was given by Philip (1969). He showed that where the inertia terms are negligible [$(\mathbf{u} \cdot \nabla)\mathbf{u} \simeq 0$], and for steady flow ($\partial \mathbf{u}/\partial t = 0$),

$$\nu_k \nabla^2 \mathbf{u} = \nabla \phi$$

Philip showed that this equation is obeyed by both Poiseuille's and Darcy's laws provided, again, the inertia terms are negligible in relation to the viscous terms (i.e., where the capillaries or pores are sufficiently small, and the flow velocity sufficiently slow).

In a one-dimensional system, Eq. (4.6) takes the form

$$\mathbf{q} = -K\frac{dH}{dx} \tag{4.7}$$

Mathematically, Darcy's law is similar to the linear transport equations of classical physics, including *Ohm's law* (which states that the current, or flow rate of electricity, is proportional to the electrical potential gradient); *Fourier's law* (the rate of heat conduction is proportional to the temperature gradient); and *Fick's law* (the rate of diffusion is proportional to the concentration gradient).

C. Gravitational, Pressure, and Total Hydraulic Heads

The water entering the column of Fig. 4.3 is under a pressure P_i, which is the sum of the hydrostatic pressure P_s and the atmospheric pressure P_a acting on the surface of the water in the reservoir. Since the atmospheric pressure is the same at both ends of the system, we can disregard it, and consider only the hydrostatic pressure. Accordingly, the water pressure at the inflow boundary is $\rho_w g H_{pi}$. Since ρ_w and g are both nearly constant, we can express this pressure in terms of the pressure head, H_{pi}.

Water flow in a horizontal column occurs in response to a pressure-head gradient. Flow in a vertical column can be caused by gravitation as well as pressure. The gravitational head H_g at any point is determined by the height of the point relative to some reference plane, while the pressure head is determined by the height of the water column resting on that point.

The total hydraulic head H is composed of the sum of these two heads:

$$H = H_p + H_g \tag{4.8}$$

To apply Darcy's law to vertical flow, we must consider the total hydraulic head at the inflow and at the outflow boundaries (H_i and H_o, respectively):

$$H_i = H_{pi} + H_{gi}, \qquad H_o = H_{po} + H_{go}$$

Darcy's law thus becomes

$$\mathbf{q} = K\frac{(H_{pi} + H_{gi}) - (H_{po} + H_{go})}{L}$$

The gravitational head is often designated as z, which is the vertical distance in the rectangular coordinate system x, y, z. It is convenient to set the reference level $z = 0$ at the bottom of a vertical column, or at the center of a horizontal column. However, the exact elevation of this hypothetical level is unimportant, since the absolute values of the hydraulic heads determined in reference to it are immaterial and only their differences from one point in the soil to another affect flow.

The pressure and gravity heads can be represented graphically in a simple way. To illustrate this, we shall immerse and equilibrate a vertical soil column in a water reservoir, so that the upper surface of the column will be level with the water surface, as shown in Fig. 4.4.

The coordinates of Fig. 4.4 are arranged so that the height above the bottom of the column is indicated by the vertical axis z; and the pressure, gravity, and hydraulic heads are indicated on the horizontal axis. The gravity

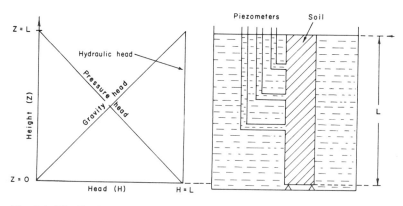

Fig. 4.4. Distribution of pressure, gravity, and total hydraulic heads in a vertical column immersed in water, at equilibrium.

head is determined with reference to the reference level $z = 0$, and increases with height at the ratio of $1 : 1$. The pressure head is determined with reference to the free water surface, at which the hydrostatic pressure is zero. Accordingly, the hydrostatic pressure head at the top of the column is zero, and at the bottom of the column it is equal to L, the column length. Just as the gravity head decreases from top to bottom, so the pressure head increases; thus, their sum, which is the hydraulic head, remains constant all along the column. This is a state of equilibrium in which no flow occurs.

This statement should be further elaborated. The water pressure is not equal along the column, as it is greater at the bottom than at the top of the column. Why, then, will not the water flow from a zone of higher to one of lower pressure? If the pressure gradient were the only force causing flow (as it is, in fact, in a horizontal column), the water would tend to flow upward. However, opposing the pressure gradient is a gravitational gradient of equal magnitude, resulting from the fact that the water at the top is at a higher gravitational potential than at the bottom. Since these two opposing gradients cancel each other, the total hydraulic head is constant, as indicated by the standpipes connected to the column at the left.

As we have already pointed out, the reference level is generally set at the bottom of the column, so that the gravitational potential can always be positive. On the other hand, the pressure head of water, which is positive under a free water surface, can in other cases be "negative." A "negative" hydraulic head signifies a pressure smaller than atmospheric. Such sub-pressures can occur above the water table and when the soil is unsaturated. Flow under these conditions will be dealt with in the next chapter.

D. Flow in a Vertical Column

Figure 4.5 shows a uniform, saturated vertical column, the upper surface of which is ponded under a constant head of water H_1, and the lower surface

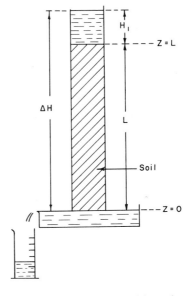

Fig. 4.5. Downward flow of water in a vertical, saturated column.

of which is set in a lower, constant-level reservoir. Flow is thus taking place from the higher to the lower reservoir through a column of length L.[5]

In order to calculate the flux according to Darcy's law, we must know the hydraulic-head gradient, which is the ratio of the hydraulic-head drop (between the inflow and outflow boundaries) to the column length:

[5] This is the same system that Darcy considered in his classic filter-bed analysis.

		Pressure head		Gravity head
Hydraulic head at inflow boundary	H_i =	H_1	+	L
Hydraulic head at outflow boundary	H_o =	0	+	0

Hydraulic head difference	$\Delta H = H_i - H_o$ =	H_1	+	L

The Darcy equation for this case is

$$q = K\frac{\Delta H}{L} = K\frac{H_1 + L}{L}$$

$$q = K\frac{H_1}{L} + K \tag{4.9}$$

Comparison of this case with the horizontal one shows that the rate of downward flow of water in a vertical column is greater than in a horizontal column by the magnitude of the hydraulic conductivity. It is also apparent that, if the ponding depth H_1 is negligible, the flux is equal to the hydraulic conductivity. This is due to the fact that, in the absence of a pressure gradient, the only driving force is the gravitational head gradient, which, in a vertical column, has the value of unity (since this head varies with height at the ratio of 1:1).

We shall now examine the case of upward flow in a vertical column, as shown in Fig. 4.6.

In this case, the direction of flow is opposite to the direction of the gravitational gradient, and the hydraulic gradient becomes

		Pressure head		Gravity head
Hydraulic head at inflow	H_i =	H_1	+	0
Hydraulic head at outflow	H_o =	0	+	L

$\Delta H = H_i - H_o$ =	H_1	-	L

The Darcy equation is thus

$$q = K\frac{H_1 - L}{L} = K\frac{H_1}{L} - K$$

$$q = K\frac{\Delta H}{L}$$

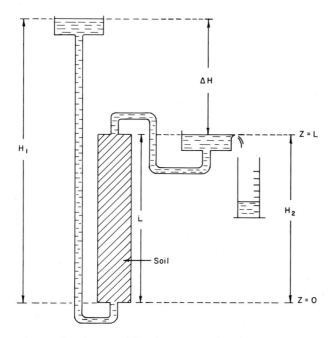

Fig. 4.6. Steady upward flow in a saturated vertical column.

E. Flow in a Composite Column

We shall consider briefly a nonuniform soil column consisting of two distinct layers, each homogeneous within itself and characterized by its own thickness and hydraulic conductivity. Assume that layer 1 is at the inlet and layer 2 at the outlet side of the flowing column, and that H_1 is the hydraulic head at the inlet surface, H_2 at the interlayer boundary, and H_3 at the outlet end. At steady flow, the flux through both layers must be equal:

$$q = K_1 \frac{(H_1 - H_2)}{L_1} = K_2 \frac{(H_2 - H_3)}{L_2} \tag{4.10}$$

where q is the flux, K_1 and L_1 are the conductivity and thickness (respectively) of the first layer, and K_2 and L_2 the same for the second layer. Here, we have disregarded any possible contact resistance between the layers. Thus,

$$H_2 = H_1 - q \frac{L_1}{K_1}$$

and

$$q \frac{L_2}{K_2} = H_2 - H_3$$

Therefore,

$$q\frac{L_2}{K_2} = H_1 - q\frac{L_1}{K_1} - H_3$$

$$q = \frac{H_1 - H_3}{L_2/K_2 + L_1/K_1} \qquad (4.11)$$

The reciprocal of the conductivity has been called the *hydraulic resistivity*. and the ratio of the thickness to the conductivity ($R_s = L/K$) has been called the *hydraulic resistance* per unit area (Hillel and Gardner, 1969). Hence,

$$q = \frac{\Delta H}{R_{s_1} + R_{s_2}} \qquad (4.12)$$

where ΔH is the total hydraulic-head drop across the entire system and R_{s_1} and R_{s_2} are the resistances of layers 1 and 2. Equation (4.12) is in complete analogy of Ohm's law for constant resistances in series.

F. Flux, Flow Velocity, and Tortuosity

As stated above, the flux density (hereafter, simply "flux") is the volume of water passing through a unit cross-sectional area (perpendicular to the flow direction) per unit time. The dimensions of the flux are:

$$q = V/At = L^3/L^2T = LT^{-1}$$

i.e., length per time (in cgs units, centimeters per second). These are the dimensions of velocity, yet we prefer the term "flux" to "flow velocity," the latter being an ambiguous term. Since soil pores vary in shape, width and direction, the actual flow velocity in the soil is highly variable (e.g., wider pores conduct water more rapidly, and the liquid in the center of each pore moves faster than the liquid in close proximity to the particles). Strictly speaking, therefore, one cannot refer to a single velocity of liquid flow, but at best of an average velocity.

Yet, even the average velocity of the flowing liquid differs from the flux, as we have defined it. Flow does not in fact take place through the entire cross-sectional area A, since part of this area is plugged by particles and only the porosity fraction is open to flow. Since the real area is smaller than A, the actual average velocity of the liquid must be greater than the flux q. Furthermore, the actual flow path is greater than the soil column length L, owing to tortuosity, as shown in Fig. 4.7.

The *tortuosity* can be defined as the average ratio of the actual "roundabout" path to the apparent, or "straight" flow path; i.e., it is the ratio of

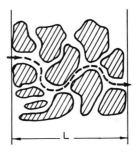

Fig. 4.7. Flow-path tortuosity in the soil.

the distance traversed by an "average" parcel of water flowing through the pores of a soil specimen to the length of that specimen. The tortuosity is thus a dimensionless geometric parameter of porous media which, though difficult to measure precisely, is always greater than 1 and may exceed 2 (Scheidegger, 1957). The *tortuosity factor* is sometimes defined as the inverse of the above.

G. Hydraulic Conductivity, Permeability, and Fluidity

The hydraulic conductivity, again, is the ratio of the flux to the hydraulic gradient, or the slope of the flux vs. gradient curve (Fig. 4.8).

With the dimensions of flux being LT^{-1}, the dimensions of hydraulic conductivity depend on the dimensions of the driving force (the potential gradient). In the last chapter, we showed that the simplest way to express the

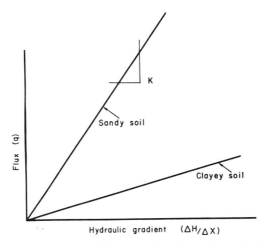

Fig. 4.8. The linear dependence of flux upon hyraulic gradient, the hydraulic conductivity being the slope (i.e., the flux per unit gradient).

potential gradient is by use of length, or head, units. The hydraulic-head gradient $\Delta H/L$, being the ratio of a length to a length, is dimensionless.[6] Accordingly, the dimensions of the hydraulic conductivity are the same as the dimensions of the flux, namely LT^{-1}. If, on the other hand, the hydraulic gradient is expressed in terms of the variation of pressure with length, then the hydraulic conductivity assumes the dimensions of $M^{-1}L^3T$. Since the latter is cumbersome, the use of head units is generally preferred.

In a saturated soil of stable structure, as well as in a rigid, porous medium such as sandstone, for instance, the hydraulic conductivity is characteristically constant. Its order of magnitude is about 10^{-2}–10^{-3} cm/sec in a sandy soil, and 10^{-4}–10^{-7} cm/sec in a clayey soil.

The hydraulic conductivity is obviously affected by structure as well as by texture, being greater if the soil is highly porous, fractured, or aggregated than if it is tightly compacted and dense. The conductivity depends not only on the total porosity, but also, and primarily, on the sizes of the conducting pores. For example, a gravelly or sandy soil with large pores can have a conductivity much greater than a clay soil with narrow pores though the total porosity of a clay is generally greater than that of a sandy soil. Cracks, worm holes, and decayed root channels are present in the field and may affect flow in different ways, depending on the direction and condition of the flow process. If the pressure head is positive, these passages will run full of water and contribute greatly to the observed flux and measured conductivity. If the pressure head in the water is negative, that is, if soil water is under suction, large cavities will generally be drained and fail to transmit water.

In many soils, the hydraulic conductivity does not in fact remain constant. Because of various chemical, physical and biological processes, the hydraulic conductivity may change as water permeates and flows in a soil. Changes occurring in the composition of the exchangeable-ion complex, as when the water entering the soil has a different concentration of solutes than the original soil solution, can greatly change the hydraulic conductivity (Reeve et al., 1954; Quirk and Schofield, 1955; Brooks et al., 1956). In general, the conductivity decreases with decreasing concentration of electrolytic solutes (Reeve, 1957), due to swelling and dispersion phenomena, which are also affected by the species of cations present. The detachment and migration of clay particles during prolonged flow may result in the clogging of pores.

In practice, it is extremely difficult to saturate a soil with water without

---

[6] Though, strictly speaking, H is not a true length, but a pressure-equivalent in terms of a water-column height; $H = P/\rho_w g$, and its gradient should be assigned the units of cm_{H_2O}/cm.

Fig. 4.9. An entrapped air bubble plugging flow.

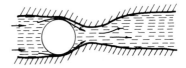

trapping some air. Entrapped air bubbles may block pore passages, as shown in Fig. 4.9. Temperature changes may cause the flowing water to dissolve or to release gas, and will also cause a change in the volume of the gas phase, thus affecting conductivity.

The hydraulic conductivity K is not an exclusive property of the soil alone, since it depends upon the attributes of the soil and of the fluid together. The soil characteristics which affect K are the total porosity, the distribution of pore sizes, and the tortuosity—in short, the pore geometry of the soil. The fluid attributes which affect the conductivity are fluid density and viscosity.

It is possible in theory, and sometimes in practice, to separate K into two factors: *intrinsic permeability* of the soil k and *fluidity* of the fluid f:

$$K = kf \qquad (4.13)$$

When K is expressed in terms of cm/sec (LT^{-1}), k is expressed in cm^2 (L^2) and f in 1/cm-sec $(L^{-1}T^{-1})$.

Fluidity is inversely proportional to viscosity:

$$f = \frac{\rho g}{v} \qquad (4.14)$$

hence,

$$k = \frac{Kv}{\rho g} \qquad (4.15)$$

where v is the viscosity in poise units (dyne sec/cm^2), ρ is the fluid density (gm/cm^3), and g is the gravitational acceleration (cm/sec^2).

In an ordinary liquid, the density is nearly constant, and changes in fluidity are likely to result primarily from changes in viscosity. In compressible fluids such as gases, on the other hand, changes in density due to pressure and temperature variation can also be considerable.

The use of the term "permeability" has in the past been a source of some confusion, as it has often been applied as synonymous or alternative to hydraulic conductivity. Permeability has also been used in a loosely qualitative sense to describe the readiness with which a porous medium transmits water or various other fluids. For this reason, the use of permeability in a strict, quantitative sense with the dimensions of length-squared [as defined above, Eq. 4.15] may require the use of some such qualifying adjective as "intrinsic" permeability (Richards, 1954) or "specific" permeability (Scheidegger, 1957). For convenience, however, we shall henceforth refer to k simply as "permeability."

It should be clear from the foregoing that, while fluidity varies with composition of the fluid and with temperature, the permeability is ideally an exclusive property of the porous medium and its pore geometry alone—provided the fluid and the solid matrix do not interact in such a way as to change the properties of either. In a completely stable porous body, the same permeability will be obtained with different fluids, e.g., with water, air, or oil. In many soils, however, matrix–water interactions are such that conductivity cannot be resolved into separate and exclusive properties of water and of soil, and Eq. (4.13) is impractical to apply.

H. Limitations of Darcy's Law

Darcy's law is not universally valid for all conditions of liquid flow in porous media. It has long been recognized that the linearity of the flux vs. hydraulic gradient relationship fails at high flow velocities, where inertial forces are no longer negligible compared to viscous forces (Hubbert, 1956). Darcy's law applies only as long as flow is laminar (i.e., non turbulent slippage of parallel layers of the fluid one atop another), and where soil–water interaction does not result in a change of fluidity or of permeability with a change in gradient. Laminar flow prevails in silts and finer materials for any commonly occurring hydraulic gradients found in nature (Klute, 1965). In coarse sands and gravels, however, hydraulic gradients much in excess of unity may result in nonlaminar flow conditions, and Darcy's law may not always be applicable.

The quantitative criterion for the onset of turbulent flow is the *Reynolds number* R_e:

$$R_e = \frac{d\bar{u}\rho}{v} \qquad (4.16)$$

where $\bar{u}$ is the mean flow velocity, d the effective pore diameter, ρ the liquid density, and v its viscosity. In straight tubes, the critical value of R_e beyond which turbulence sets in has variously been reported to be of the order of 1000–2200 (Scheidegger, 1957; Childs, 1969). However, the critical Reynolds number at which water flowing in a tube becomes turbulent is apparently reduced greatly when the tube is curved. For porous media, therefore, it is safe to assume that flux remains linear with hydraulic gradient only as long as R_e is smaller than unity. As flow velocity increases, especially in systems of large pores, the occurrence of turbulent eddies or nonlinear laminar flow results in "waste" of effective energy; i.e., the hydraulic potential gradient becomes less effective in inducing flow. This is illustrated in Fig. 4.10.

▢/Deviations from Darcy's law may also occur at the opposite end of the flow-velocity range, namely at low gradients and in small pores. Some investigators (Swartzendruber, 1962; Miller and Low, 1963; Nerpin *et al.*, 1966) have claimed that, in clayey soils, low hydraulic gradients may cause no flow or only low flow rates that are less than proportional to the gradient, while others have disputed some of these findings (Olsen, 1965). A possible reason for this anomaly is that the water in close proximity to the particles

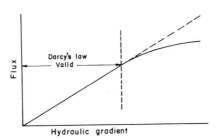

Fig. 4.10. The deviation from Darcy's law at high flux, where flow becomes turbulent.

and subject to their adsorptive force fields may be more rigid than ordinary water, and exhibit the properties of a "Bingham liquid" (having a yield value) rather than a "Newtonian liquid." The adsorbed, or bound, water may have a quasicrystalline structure similar to that of ice, or even a totally different structure.[7] Some soils may exhibit an apparent "threshold gradient," below which the flux is either zero (the water remaining apparently immobile), or at least lower than predicted by the Darcy relation,[8] and only at gradients exceeding the threshold value does the flux become proportional to the gradient (Fig. 4.11). These phenomena and their possible explanation, though highly interesting, are generally of little or no importance in practice, and Darcy's law can be employed in the vast majority of cases concerning the flow of water in soil.

---

[7] It has been speculated (Low, 1970) that the structure of water adsorbed on clay may be related to the occurrence of a heretofore unrecognized form of polymerized water, as described by Deryaguin *et al.* (1965). A recent report on this phenomenon, variously called "anomolous water," "superwater," or "polywater," is by Lippincot *et al.* (1969).

[8] Another possible cause for apparent flow anomalies in clay soils is their compressibility. Darcy's law applies to flow relative to the particle, and it may seem to fail when the particles themselves are moving relative to a fixed frame of reference.

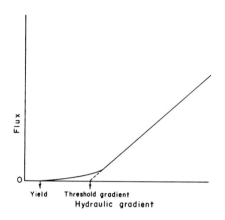

Fig. 4.11. Possible deviations from Darcy's law at low gradients.

I. Relation of Conductivity and Permeability to Pore Geometry

Since permeability is a characteristic physical property of a porous medium, it would seem only reasonable to assume that it relates in some functional way to certain measurable properties of the soil pore geometry, e.g., porosity, pore-size distribution, internal surface area, etc. However, numerous attempts to discover a functional relation of universal applicability have so far met with disappointing results.

Perhaps the simplest approach is to seek a correlation between permeability and porosity (e.g., Jacob, 1946; Franzini, 1951). The reader should have concluded by now, however, that this is, in general, a futile approach (except for the comparison of otherwise identical media), owing to the strong dependence of flow rate upon the width, continuity, shape, and tortuosity of the conducting channels. Thus, a medium composed of numerous small pores with a high total porosity is likely to exhibit a lower saturated conductivity than a medium of lesser porosity but larger (though fewer) individual pores.

Attempts have also been made to find correlations between permeability and grain-size distribution (Pillsbury, 1950). Such correlations may indeed be found for similar materials, but hardly for materials of different grain shapes and aggregations (e.g., sand vs. clay). Attempts to refine this approach by introducing empirical grain-shape and packing parameters (Tickell and Hyatt, 1938) have not won general acceptance.

Numerous attempts have been made to represent porous media by idealized theoretical models which are amenable to mathematical treatment. Some of these models are highly elegant, yet their worth must depend on experiment, which alone can show whether or not they truly represent the behavior of real porous media. Scheidegger (1957) gave a comprehensive review of such models, including the "straight capillaric," "parallel," "serial," and "branching" models. He pointed out that, in general, natural porous media are extremely disordered, so that it seems a rather poor procedure to represent them by something which is intrinsically ordered.[9] He therefore suggested that the preferred model of a porous medium should be based upon statistical concepts.

One of the most widely accepted theories on the relation of permeability to the geometric properties of porous media is the Kozeny theory, and particularly its modification by Carman (1939). This theory is based on the concept of a "hydraulic radius," i.e., a characteristic length parameter presumed to be linked with the hypothetical channels to which the porous medium is thought to be equivalent. The measure of the hydraulic radius is the ratio of the volume to the surface of the pore space, or the average ratio of the cross-sectional area of the pores to their circumferences. The following is known as the Kozeny–Carman equation:

$$k = \frac{f^3}{ca^2(1-f)^2} \tag{4.17}$$

where f is the porosity, a the specific surface exposed to the fluid, and c a constant representing a particle shape factor. For a critique of this theory, see Scheidegger (1957). A criticism particularly apt in relation to soils is that the hydraulic-radius theory fails to describe structured bodies such as, for example, fissured clays, where the structural fissures contribute negligibly both to porosity and specific surface, and yet they dominate the permeability (Childs and Collis-George, 1950).

A promising approach to the prediction of permeability from basic physical properties of the porous medium is to seek a connection between permeability and pore-size distribution. Since, however, there is no direct or simple way to obtain or characterize this distribution *per se*, it is only possible to work with parameters which are based indirectly upon the pore-size distribution, namely parameters based on the suction or capillary pressure vs. sorption or desorption. Since flow through an irregular pore is limited by the narrow "necks" along the flow paths, one needs also to consider or estimate the number and size of "necks" and the interconnections of pores of different widths. Work along these lines has been published by Childs and

---

[9] An ordered medium consists of a sequence of internally identical units having some consistent geometric pattern.

Collis-George (1950), Marshall (1958),[10] and Millington and Quirk (1959). The results of these theories, while more generally applicable than those based on earlier models, still appear to be valid only for certain coarse materials in which capillary phenomena predominate.

J. Homogeneity and Isotropy

The hydraulic conductivity (or permeability) may be uniform throughout the soil, or may vary from point to point, in which case the soil is said to be hydraulically *inhomogeneous*. If the conductivity is the same in all directions, the soil is *isotropic*. However, the conductivity at each point may differ for different directions (e.g., the horizontal conductivity may be greater, or smaller, than the vertical), a condition known as *anisotropy*. A soil may be homogeneous and nevertheless anisotropic, or it may be inhomogeneous (e.g., layered) and yet isotropic at each point. Some soils exhibit both in-homogeneity and anisotropy. In certain cases, K may also be *asymmetrical*, that is to say, indicate a different value depending on the direction of flow along a given line. The measurement of the directional permeability of soils was discussed by Measland and Kirkham (1955). A comprehensive review of anisotropy and layering is given in the book by Bear *et al.* (1968). Anisotropy is generally due to the structure of the soil, which may be laminar, or platy, or columnar, etc., thus exhibiting a pattern of micropores or macropores with a distinct directionality.

K. Measurement of Hydraulic Conductivity of Saturated Soils

Methods for measuring hydraulic conductivity in the laboratory were reviewed recently by Klute (1965), and for measurement in the field by Talsma (1960) and by Boersma (1965). The use of permeameters for laboratory determinations is illustrated in Figs. 4.12 and 4.13. Such determinations can be made with dried and fragmented specimens, which then must be packed into the flow cells in a standard manner, or, preferably, with undisturbed

[10] Marshall's approach is based on dividing the soil-moisture characteristic curve into equally-spaced segments, each characterized by a value of matric suction sufficient to empty it (assuming the capillary equation $P = -2\gamma/r$). His equation is

$$K = \frac{\varepsilon^2}{8n^2} [r_1{}^2 + 3r_2{}^2 + 5r_3{}^2 + \cdots + (2n-1)r_n{}^2]$$

where ε is porosity and r_1, r_2, and r_n represent the mean radii of the pores (in decreasing order of size) in each of the n equal fractions of the total pore space.

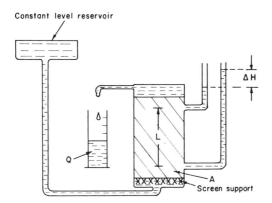

Fig. 4.12. The measurement of saturated hydraulic conductivity with a constant-head permeameter; $K = VL/At\ \Delta H$.

core samples taken directly from the field. In either case, provision must be made to avoid boundary flow along the walls of the container. Field measurements can be made most conveniently below the water table, as by the auger-hole method (Luthin, 1957) or by the piezometer method (Johnson *et al.*, 1952). Techniques have also been proposed for measurements above the

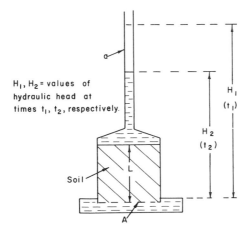

Fig. 4.13. The measurement of saturated hydraulic conductivity with a falling-head permeameter; $K = [2.3aL/A(t_2 - t_1)](\log H_1 - \log H_2)$. H_1 and H_2 are the values of hydraulic head at times t_1 and t_2, respectively.

water table, as by the double-tube method (Bouwer, 1961, 1962), the shallow-well pump-in method, and the field-permeameter method (Winger, 1960).

L. Equations of Saturated Flow

Darcy's law, by itself, is sufficient only to describe steady, or stationary, flow processes, in which the flux remains constant and equal along the conducting system (and hence the potential and gradient at each point remain constant with time). Unsteady, or transient, flow processes, in which the magnitude and possibly even the direction of the flux and potential gradient vary with time, require the introduction of an additional law, namely, the law of conservation of matter. To understand how this law applies to flow phenomena, consider a small volume element (say, a cube) of soil, into and out of which flow takes place at possibly differing rates. The mass-conservation law, expressed in the *equation of continuity*, states that if the rate of inflow into the volume element is greater than the rate of outflow, then the volume element must be storing the excess and increasing its water content (and conversely, if outflow exceeds inflow, storage must decrease).

Considering first the simplest case of one-dimensional flow, with q_x being the flux in the x direction, the rate of increase of q_x with x must equal the rate of decrease of volumetric water content θ with time t:

$$\frac{\partial \theta}{\partial t} = -\frac{\partial q_x}{\partial x} \tag{4.18}$$

which, in multidimensional systems, becomes

$$\frac{\partial \theta}{\partial t} = -\nabla \cdot \mathbf{q} \tag{4.19}$$

We recall Darcy's law,

$$\mathbf{q} = -K\nabla H \tag{4.20}$$

which, in one dimension, is

$$q_x = -K\frac{dH}{dx} \tag{4.21}$$

(where H is the hydraulic head and K the hydraulic conductivity). Now, we combine this with the continuity equation (4.19) to obtain the general flow equation:

$$\frac{\partial \theta}{\partial t} = \nabla \cdot K\nabla H \tag{4.22}$$

In applying this equation, the assumptions are usually made that inertial forces are negligible in comparison with viscous forces, that the water is continuously connected throughout the flow region, that isothermal conditions prevail, and that no chemical or biological phenomena change the fluid or the porous medium.

In one dimension, Eq. (4.22) becomes

$$\frac{\partial \theta}{\partial t} = \frac{\partial}{\partial x}\left(K\frac{\partial H}{\partial x}\right) \tag{4.23}$$

since the hydraulic head can be resolved into a pressure head H_p and a gravitational head (an elevation above some reference datum, z) we can write

$$\frac{\partial \theta}{\partial t} = \text{div}[K(\nabla H_p + \nabla z)] \tag{4.24}$$

In horizontal flow, $\nabla z = 0$, so for this case,

$$\frac{\partial \theta}{\partial t} = \text{div}K\left(\frac{\partial H_p}{\partial x}\right) \tag{4.25}$$

while in vertical flow, $\nabla z = 1$, and therefore, for this case,

$$\frac{\partial \theta}{\partial t} = \text{div}\left[K\left(\frac{\partial H_p}{\partial z} + 1\right)\right] \tag{4.26}$$

In a saturated soil with an incompressible matrix, $\partial \theta / \partial t = 0$, the conductivity is usually assumed to remain constant, hence Eq. (4.23) becomes

$$K_s \frac{\partial^2 H}{\partial x^2} = 0 \tag{4.27}$$

where K_s is the hydraulic conductivity of the saturated soil (the "saturated conductivity"). For three-dimensional flow conditions, and allowing for anisotropy, the equation is

$$K_x \frac{\partial^2 H}{\partial x^2} + K_y \frac{\partial^2 H}{\partial y^2} + K_z \frac{\partial^2 H}{\partial z^2} = 0 \tag{4.28}$$

where K_x, K_y, and K_z represent the hydraulic conductivity values in the three principal directions x, y, z.

In an isotropic soil (where $K_x = K_y = K_z$ at each point) that is also homogeneous (K equal at all points), we obtain the well-known *Laplace equation*:

$$\frac{\partial^2 H}{\partial x^2} + \frac{\partial^2 H}{\partial y^2} + \frac{\partial^2 H}{\partial z^2} = 0 \tag{4.29}$$

This is a second-order partial differential equation of the elliptical type, and it can be solved in certain cases to obtain a quantitative description of water flow in various systems.

In general, a differential equation can have an infinite number of solutions. To determine the specific solution in any given case, it is necessary to specify the *boundary conditions*, and, in the case of unsteady flow, of the *initial conditions* as well. Various types of boundary conditions can exist (e.g., impervious boundaries, free water surfaces, boundaries of known pressure, or known inflow or outflow rates, etc.), but in each case the flux and pressure head must be continuous throughout the system. In layered soils, the hydraulic conductivity and water content may be discontinuous across interlayer boundaries (that is, they may exhibit abrupt changes). Flow equations for inhomogeneous, anisotropic, and compressible systems were given by Bear *et al.* (1968).

Philip (1969) recently analyzed flow in swelling (compressible) media. In unsteady flow, the solid matrix of a swelling soil undergoes motion, so that Darcy's law applies to water movement relative to the particles, rather than relative to physical space. Experimental work with such soils was carried out by Smiles and Rosenthal (1968).

M. Summary

A proper physical description of water flow in the soil requires that three parameters be specified: flux, hydraulic gradient, and conductivity. Knowledge of any two of these allows the calculation of the third, according to Darcy's law. This law states that the flux equals the product of conductivity by the hydraulic gradient. The hydraulic gradient itself includes both the pressure and the gravitational potential gradients, the first of which is the exclusive cause of flow in a horizontal system, while the second occurs in vertical systems. The hydraulic conductivity at saturation is a characteristic property of a soil toward water flow, and it is related to porosity and pore-size distribution.

5 *Flow of Water in Unsaturated Soil*

A. General

Most of the processes involving soil-water flow in the field, and in the rooting zone of most plant habitats, occur while the soil is in an unsaturated condition. Unsaturated flow processes are in general complicated and difficult to describe quantitatively, since they often entail changes in the state and content of soil water during flow. Such changes involve complex relations among the variable water content (wetness), suction, and conductivity, which may be affected by hysteresis. The formulation and solution of unsaturated flow problems very often require the use of indirect methods of analysis, based on approximations or numerical techniques. For this reason, the development of rigorous theory and methods for treating these problems was rather late in coming. In recent years, however, unsaturated flow has become one of the most important and active topics of research in soil physics, and this research has resulted in significant theoretical and practical advances.

B. Comparison of Unsaturated vs. Saturated Flow

In the previous chapter, we stated that soil-water flow is caused by a driving force resulting from an effective potential gradient, that flow takes place in the direction of decreasing potential, and that the rate of flow (flux) is proportional to the potential gradient and is affected by the geometric properties of the pore channels through which flow takes place. These principles apply in unsaturated, as well as in saturated soils.

The moving force in a saturated soil is the gradient of a positive pressure potential.[1] On the other hand, water in an unsaturated soil is subject to a subatmospheric pressure, or suction, and the gradient of this suction likewise constitutes a moving force. The matric suction is due, as we have pointed out, to the physical affinity of the water to the soil-particle surfaces and capillary pores. Water tends to be drawn from a zone where the hydration envelopes surrounding the particles are thicker, to where they are thinner, and from a zone where the capillary menisci are less curved to where they are more highly curved.[2] In other words, water tends to flow from where suction is low to where it is high. When suction is uniform all along a horizontal column, that column is at equilibrium and there is no moving force. Not so when a suction gradient exists. In that case, water will flow in the pores which remain water-filled at the existing suction, and will creep along the hydration films over the particle surfaces, in a tendency to equilibrate the potential.

The moving force is greatest at the "wetting front" zone of water entry into an originally dry soil (see Fig. 5.2). In this zone, the suction gradient can be many bars per centimeter of soil. Such a gradient constitutes a moving force thousands of times greater than the gravitational force. As we shall see later on, such strong forces are sometimes required (for a given flux) in view of the extremely low hydraulic conductivity which a relatively dry soil may exhibit.

The most important difference between unsaturated and saturated flow is in the hydraulic conductivity. When the soil is saturated, all of the pores are filled and conducting, so that conductivity is maximal. When the soil becomes unsaturated, some of the pores become airfilled and the conductive portion of the soil's cross-sectional area decreases correspondingly. Furthermore, as suction develops, the first pores to empty are the largest ones, which

---

[1] We shall disregard, for the moment, the gravitational force, which is completely unaffected by the saturation or unsaturation of the soil.

[2] The question of how water-to-air interfaces behave in a conducting porous medium that is unsaturated is imperfectly understood. It is generally assumed, at least implicitly, that these interfaces, or menisci, are anchored rigidly to the solid matrix so that, as far as the flowing water is concerned, air-filled pores are like solid particles. The presence of organic surfactants which adsorb to these surfaces is considered to increase their rigidity or viscosity. Even if the air–water interfaces are not entirely stationary, however, the drag, or momentum transfer, between flowing water and air appears to be very small. The influence of the surface viscosity of air–water interfaces on the rheological behavior of soil water has not been evaluated (Philip, 1970). Preliminary experimental findings by E. E. Miller and D. Hillel suggest that a drag effect does occur, but that its magnitude is negligible for most practical purposes.

are the most conductive,[3] thus leaving water to flow only in the smaller pores. The empty pores must be circumvented, so that, with desaturation, the tortuosity increases. In coarse-textured soils, water sometimes remains almost entirely in capillary wedges at the contact points of the particles, thus forming separate and discontinuous pockets of water. In aggregated soils, too, the large interaggregate spaces which confer high conductivity at saturation become (when emptied) barriers to liquid flow from one aggregate to its neighbors.

For these reasons, the transition from saturation to unsaturation generally entails a steep drop in hydraulic conductivity, which may decrease by several orders of magnitude (sometimes down to 1/100,000 of its value at saturation) as suction increases from zero to one bar. At still higher suctions, or lower water contents, the conductivity may be so low[4] that very steep suction gradients, or very long times, are required for any appreciable flow to occur.

At saturation, the most conductive soils are those in which large and continuous pores constitute most of the overall pore volume, while the least conductive are the soils in which the pore volume consists of numerous micropores. Thus, as is well known, a sandy soil conducts water more rapidly than a clayey soil. However, the very opposite may be true when the soils are unsaturated. In a soil with large pores, these pores quickly empty and become nonconductive as suction develops, thus steeply decreasing the initially high conductivity. In a soil with small pores, on the other hand, many of the pores remain full and conductive even at appreciable suction, so that the hydraulic conductivity does not decrease as steeply and may actually be greater than that of a soil with large pores subjected to the same suction.

Since in the field the soil is unsaturated most of the time, it often happens that flow is more appreciable and persists longer in clayey than in sandy soils. For this reason, the occurrence of a layer of sand in a fine-textured profile, far from enhancing flow, may actually impede unsaturated water movement until water accumulates above the sand and suction decreases sufficiently for water to enter the large pores of the sand. This simple principle is all too often misunderstood.

---

[3] By Poiseuille's law, the total flow rate of water through a capillary tube is proportional to the fourth power of the radius, while the flow rate per unit cross-sectional area of the tube is proportional to the square of the radius. A 1-mm-radius pore will thus conduct as 10,000 pores of radius 0.1 mm.

[4] As very high suctions develop, there may (in addition to the increase in tortuosity and the decrease in number and sizes of the conducting pores) also be a change in the viscosity of the (mainly adsorbed) water, tending to further reduce the conductivity. (Miller and Low, 1963).

C. Relation of Conductivity to Suction and Wetness

Let us consider an unsaturated soil in which water is flowing under suction. Such flow is illustrated schematically in the model of Fig. 5.1. In this model, the potential difference between the inflow and outflow ends is maintained not by different heads of positive hydrostatic pressure, but by different imposed suctions. In general, as the suction varies along the sample, so will the wetness and conductivity. If the suction heads at both ends of the sample

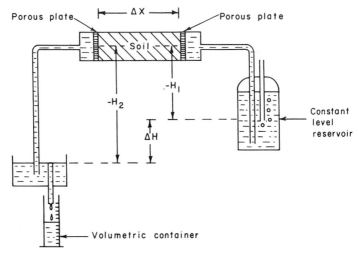

Fig. 5.1. A model illustrating unsaturated flow (under a suction gradient) in a horizontal column.

are maintained constant, the flow process will be steady and the suction gradient will increase as the conductivity decreases with the increase in suction along the axis of the sample. This phenomenon is illustrated in Fig. 5.2.

Since the gradient along the column is not constant, as it is in uniform saturated systems, it is not possible, strictly speaking, to divide the flux by the overall ratio of the head drop to the distance ($\Delta H/\Delta x$) to obtain the conductivity. Rather, it is necessary to divide the flux by the exact gradient at each point to evaluate the exact conductivity and its variation with suction. In the following treatment, however, we shall assume that the column of Fig. 5.1 is sufficiently short to allow us to evaluate at least an average conductivity for the sample as a whole (i.e., $K = q\Delta x/\Delta H$).

The average negative head, or suction, acting in the column is:

$$-\overline{H} = \overline{\psi} = -\frac{H_1 + H_2}{2}$$

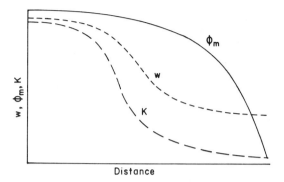

Fig. 5.2. The variation of wetness w, matric potential ϕ_m, and conductivity K along a hypothetical column of unsaturated soil conducting a steady flow of water.

We assume that the suction everywhere exceeds the air-entry value so that the soil is unsaturated throughout.

Let us now make successive and systematic measurements of flux vs. suction gradient for different values of average suction. The results of such a series of measurements are shown schematically in Fig. 5.3. As in the case of saturated flow, we find that the flux is proportional to the gradient. However, the slope of the flux vs. gradient line, being the hydraulic conductivity, varies with the average suction. In a saturated soil, by way of contrast, the hydraulic conductivity is generally independent of the magnitude of the water potential, or pressure.

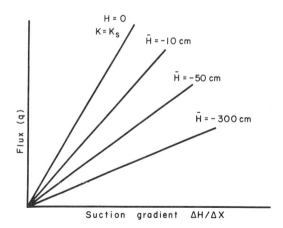

Fig. 5.3. The hydraulic conductivity, being the slope of the flux vs. gradient relation, depends upon the average suction in an unsaturated soil.

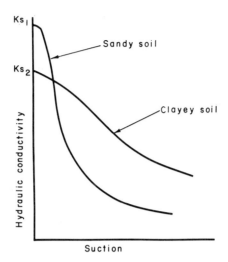

Fig. 5.4. Dependence of conductivity on suction in soils of different texture (log–log scale).

Figure 5.4 shows the general trend of the dependence of conductivity on suction[5] in soils of different texture. It is seen that, although the saturated conductivity of the sandy soil K_{s1} is typically greater than that of the clayey soil K_{s2}, the unsaturated conductivity of the former decreases more steeply with increasing suction and eventually becomes lower.

No fundamentally based equation of general validity is available for the relation of conductivity to suction or to wetness, and existing knowledge does not allow the reliable prediction of unsaturated conductivity from basic soil properties. Various empirical equations have been proposed, however, including the following (Gardner, 1960):

$$K = \frac{a}{\psi^m} \tag{5.1a}$$

$$K = \frac{a}{b + \psi^m} \tag{5.1b}$$

$$K = \frac{K_s}{1 + (\psi/\psi_c)^m} \tag{5.1c}$$

$$K = a\theta^m \tag{5.1d}$$

$$K = K_s W_s^{\ m} \tag{5.1e}$$

[5] K vs. suction curves are usually drawn on a log–log scale, as both K and ψ vary over several orders of magnitude within the suction range of general interest (say, 0 to 10,000 cm of suction head).

where K is the hydraulic conductivity at any degree of saturation (or unsaturation); K_s is the saturated conductivity of the same soil; a, b, and m are empirical constants (different in each equation); ψ is the matric suction head; θ is the volumetric water content; W_s is the degree of saturation; and ψ_c is the suction head at which $K = \frac{1}{2}K_s$.

Of these various equations, the most commonly employed are the first two (of which the first is the simplest, but cannot be used in the suction range approaching zero). In all of the equations, the most important parameter is the exponential constant, since it controls the steepness with which conductivity decreases with increasing suction or with decreasing water content. The m value of the first two equations is about two or less for clayey soils, and may be four or more for sandy soils. For each soil, the equation of best fit, and the values of the parameters, must be determined experimentally.

The relation of conductivity to suction depends upon hysteresis, and is thus different in a wetting than in a drying soil. The reason is that, at a given suction, a drying soil contains more water than a wetting one. The relation of conductivity to water content, however, appears to be affected by hysteresis to a much lesser degree (Topp and Miller, 1966; Poulovassilis, 1969). The value of the exponent for the relation of K to θ, (Eq. 5.1d), can be as high as 10 or more (Gardner et al., 1970).

Some investigators have used a different designation, usually "capillary conductivity," to distinguish the hydraulic conductivity of a soil at unsaturation from that at saturation. This, however, is generally unnecessary and the adjective "capillary" can be misleading, since unsaturated flow may not conform to the capillary model any more than does saturated flow.

D. General Equation of Unsaturated Flow

Darcy's law, though originally conceived for saturated flow only, was extended by Richards (1931) to unsaturated flow, with the provision that the conductivity is now a function of the matric suction head [i.e., $K = K(\psi)$]:

$$\mathbf{q} = - K(\psi)\, \nabla H \qquad (5.2)$$

where ∇H is the hydraulic head gradient, which may include both suction and gravitational components.

As pointed out by Miller and Miller (1956), this formulation fails to take into account the hysteresis of soil-water characteristics. In practice, the hysteresis problem can sometimes be evaded by limiting the use of Eq. (5.1) to cases in which the suction (or wetness) change is monotonic— that is, either increasing or decreasing continuously. In processes involving both wetting and drying phases, Eq. (5.1) is difficult to apply, as the $K(\psi)$ function may be highly hysteretic. As mentioned in the previous section, however, the relation of conductivity to volumetric wetness $K(\theta)$ or to degree of saturation $K(w_s)$ is affected by hysteresis to a much lesser degree than is the $K(\psi)$ function, at least in the media thus far examined (Topp and Miller, 1966). Thus, Darcy's law for unsaturated soil can also be written as

$$\mathbf{q} = -K(\theta)\,\nabla H \tag{5.3}$$

which, however, still leaves us with the problem of dealing with the hysteresis between ψ and θ.

To obtain the general flow equation and account for transient, as well as steady, flow processes, we must introduce the continuity equation

$$\frac{\partial\theta}{\partial t} = -\nabla\cdot\mathbf{q} \tag{5.4}$$

Thus,

$$\frac{\partial\theta}{\partial t} = \nabla\cdot[K(\psi)\,\nabla H] \tag{5.5}$$

Remembering that the hydraulic head is, in general, the sum of the pressure head (or its negative, the suction head) and the gravitational head (or elevation) z, we can write

$$\frac{\partial\theta}{\partial t} = -\nabla\cdot[K(\psi)\,\nabla(\psi - z)] = -\nabla\cdot(K\,\nabla\psi) + \frac{\partial K}{\partial z} \tag{5.6}$$

or

$$\frac{\partial\theta}{\partial t} = -\frac{\partial}{\partial x}\left(K\frac{\partial\psi}{\partial x}\right) - \frac{\partial}{\partial y}\left(K\frac{\partial\psi}{\partial y}\right) - \frac{\partial}{\partial z}\left(K\frac{\partial\psi}{\partial z}\right) + \frac{\partial K}{\partial z} = \frac{\partial\theta}{\partial\psi}\frac{\partial\psi}{\partial t} \tag{5.7}$$

where $d\theta/d\psi$ is the slope of the soil moisture characteristic (i.e., the specific water capacity). In horizontal flow, obviously, ∇z is zero. Other processes may also occur in which ∇z is negligible compared to the strong matric suction gradient, $\nabla\psi$. In such cases,

$$\frac{\partial\theta}{\partial t} = \nabla\cdot[K(\psi)\,\nabla\psi] \tag{5.8}$$

or, in a one-dimensional horizontal system,

$$\frac{\partial\theta}{\partial t} = \frac{\partial}{\partial x}\left[K(\psi)\frac{\partial\psi}{\partial x}\right] \tag{5.9}$$

E. Diffusivity

To simplify the mathematical and experimental treatment of unsaturated flow processes, it is often advantageous to change the flow equations into a form analogous to the equations of diffusion and heat conduction, for which ready solutions are available (e.g., Carslaw and Jaeger, 1959; Crank, 1956) in some cases involving boundary conditions applicable to soil-water flow processes. To transform the flow equation, it is sometimes possible to relate the flux to the water content (wetness) gradient rather than to the suction gradient.

The matric suction gradient $\partial \psi/\partial x$ can be expanded by the chain rule as follows

$$\frac{\partial \psi}{\partial x} = \frac{d\psi}{d\theta} \cdot \frac{\partial \theta}{\partial x} \tag{5.10}$$

where $\partial \theta/\partial x$ is the wetness gradient and $d\psi/d\theta$ is the reciprocal of the specific water capacity $c(\theta)$:

$$c(\theta) = \frac{d\theta}{d\psi} \tag{5.11}$$

which is, again, the slope of the soil-moisture characteristic curve at any particular value of wetness θ.

We can now rewrite the Darcy equation as follows:

$$q = K(\theta) \frac{\partial \psi}{\partial x} = -\frac{K(\theta)}{c(\theta)} \cdot \frac{\partial \theta}{\partial x} \tag{5.12}$$

To cast this equation into a form analogous to Fick's law of diffusion, a function was introduced (Childs and Collis-George, 1950), called the *diffusivity* D, where

$$D(\theta) = \frac{K(\theta)}{c(\theta)} = K(\theta) \frac{d\psi}{d\theta} \tag{5.13}$$

D is thus defined as the ratio of the hydraulic conductivity to the specific water capacity, and since both of these are functions of soil wetness, the diffusivity must also be so. Now we can rewrite Eq. (5.3):

$$\mathbf{q} = -D(\theta) \, \nabla \theta \tag{5.14}$$

or, in one dimension,

$$q = -D(\theta) \frac{\partial \theta}{\partial x} \tag{5.15}$$

which is mathematically identical with the first equation of diffusion. The diffusivity can thus be viewed as the ratio of the flux to the soil water content (wetness) gradient. The diffusivity D has dimensions of length squared per unit time (L^2T^{-1}), while the specific water capacity c has dimensions of volume of water per unit volume of soil per unit change in matric suction head (L^{-1}). In the use of Eq. (5.15), the gradient of wetness is taken to represent, implicitly, a gradient of matric suction, which is a true driving force.

Introducing the diffusivity into Eq. (5.9), for one-dimensional flow in the absence of gravity, we obtain[6]

$$\frac{\partial \theta}{\partial t} = \frac{\partial}{\partial x}\left[D(\theta)\frac{\partial \theta}{\partial x}\right] \tag{5.16}$$

which has only one dependent variable.[7]

In the special case that the diffusivity remains constant (though it is not generally safe to assume this except for a very small range of wetness), Eq. (5.16) can be written in the form of Fick's second diffusion equation:

$$\frac{\partial \theta}{\partial t} = D\frac{\partial^2 \theta}{\partial x^2} \tag{5.17}$$

A word of caution is now in order. In employing the diffusivity concept, and all relationships derived from it, we must remember that the process of liquid water movement in the soil is not one of diffusion but of mass flow. The borrowed term "diffusivity," if taken literally, can be misleading. Furthermore, the diffusivity equations fail wherever the hysteresis effect is appreciable or where the soil is layered, or in the presence of thermal gradients, since under such conditions flow bears no simple or consistent relation to the decreasing water-content gradient and may actually be in the opposite direction to it. On the other hand, an advantage in using the diffusivity equations is in the fact that the range of variation of diffusivity is smaller than that of conductivity.[8] Furthermore, the wetness and its gradient are

---

[6] In two dimensions, $\partial\theta/\partial t = \partial(D\,\partial\theta/\partial x)/\partial x + \partial(D\,\partial\theta/\partial y)/\partial y$; in three dimensions, $\partial\theta/t\partial = \partial(D\,\partial\theta/dx)/\partial x + \partial(D\,\partial\theta/\partial y)/\partial y + \partial(D\,\partial\theta/\partial z)/\partial z$; in two dimensions, cylindrical geometry, $\partial\theta/\partial t = (1/r)\,\partial(Dr\,\partial\theta/\partial r)/\partial r$; in three dimensions, spherical geometry, $\partial\theta/\partial t = (1/r^2)\,\partial(Dr^2\,\partial\theta/\partial r)/\partial r$.

[7] Rubin (1966) reviewed methods of reducing Eq. (5.9) to a single dependent variable.

[8] The maximum value of D found in practice is of the order of 10^4 cm^2/day. D generally decreases to about 1–10 cm^2/day at the lower limit of wetness normally encountered in the root zone. It thus varies about a thousandfold rather than about a millionfold, as does the hydraulic conductivity in the same wetness range.

often easier to measure in practice, and to relate to volume fluxes, than the suction and its gradient.

To take account of gravity [as in Eq. (5.6)], a diffusivity equation can be written in the form

$$\frac{\partial \theta}{\partial t} = \nabla \cdot [D(\theta) \, \nabla \theta] + \frac{\partial K(\theta)}{\partial z}$$

$$= \nabla \cdot [D(\theta) \, \nabla \theta] + \frac{dK}{d\theta} \frac{\partial \theta}{\partial z} \tag{5.18}$$

The relation of diffusivity to wetness is shown in Fig. 5.5. This relation is sometimes expressed in the empirical equation (Gardner and Mayhugh, 1958)

$$D(\theta) = ae^{b\theta} \tag{5.19}$$

This equation applies only to sections of the curve showing a rise in diffusivity with wetness. In the very dry range, the diffusivity often indicates an opposite trend—namely, a rise with decreasing soil wetness. This is apparently due to the contribution of vapor movement (Philip, 1955). In the very wet range, as the soil approaches complete saturation, the diffusivity becomes indeterminate as it tends to infinity (since $c(\theta)$ tends to zero).

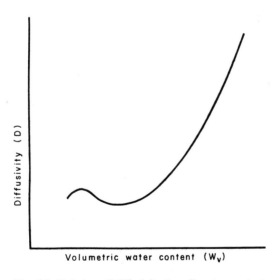

Fig. 5.5. Relation of diffusivity to soil water content.

F. The Boltzmann Solution

Because of its nonlinearity, the flow equation (5.16) for water in unsaturated soil is much more difficult to solve than are the classical linear equations for the flow of heat or electricity.

The ability of modern high-speed computers to solve nonlinear differential equations by successive numerical approximations is rapidly opening the door to practical success in broad areas of soil physics. There are also a number of simple analytical techniques that facilitate the application of the unsaturated-flow equation to particular problems. (See, for example, Gardner and Mayhugh, 1958.)

When gravity can be neglected and flow is monotonic (no hysteresis effects), the simple form $\partial\theta/\partial t = \nabla \cdot (D\,\nabla\theta)$ can be put into a form that is amenable to solution by the separation-of-variables technique so familiar in the solution of linear partial differential equations. By this means, families of one-dimensional flow patterns can be computed for Cartesian, cylindrical, and spherical coordinates. The (Cartesian) patterns conform to an important and widely studied boundary condition, namely that of a long, horizontal column of soil, initially at some uniform wetness and suction, which is suddenly subjected at one end to a different suction (either higher *or* lower). The variables-separable solution is outlined mathematically in an appendix to this book.

The variables-separable technique shows that the solution (known for its discoverer as the "Boltzmann transformation") is of the form[9]

$$B(\theta) = x/\sqrt{t} \qquad\qquad (5.20)$$

where the function $B(\theta)$ is constrained by an ordinary differential equation. Writing it in the form, θ a function of B, this ordinary differential equation for the Cartesian case becomes

$$\frac{B}{2}\frac{d\theta}{dB} = -\frac{d}{dB}\left[D(\theta)\frac{d\theta}{dB}\right] \qquad\qquad (5.21)$$

---

[9] One can perhaps perceive intuitively why the composite variable $x/\sqrt{t}$ should apply to Eq. (5.16), since this equation is first-order with respect to t and second-order with respect to x.

A common error is to regard this transformation—i.e., that θ is a simple function of the combined variable $(x/\sqrt{t})$—as a testable *assumption* pertaining to the behavior of soils. It is not. It is simply a *mathematical consequence* of the form of the differential equation. When an actual experiment fails to conform accurately to the $x/\sqrt{t}$ relation, the discrepancy can only be attributed to an imperfect description of the behavior of the soil system by the assumed differential equation and/or its assumed boundary conditions, or to errors of the experiment.

In this form, the $\theta(x, t)$ boundary conditions for horizontal infiltration can be fitted directly to the ordinary differential equations as $\theta(B)$ conditions. (θ_0 for $x = 0$, $t > 0$, becomes θ_0 for $B = 0$; and θ_i for $x > 0$, $t \to 0$, becomes θ_i for $B \to \infty$.)

This technique has been used by several investigators (Klute, 1952; Philip, 1955, 1957) to obtain solutions for soil-water flow problems that involve horizontal, semi-infinite, homogeneous media of uniform initial wetness.

G. Measurement of Unsaturated Conductivity and Diffusivity

Knowledge of the unsaturated hydraulic conductivity and diffusivity values at different suctions and water contents is generally required before any of the mathematical theories of water flow can be applied in practice. Since there is no reliable way to predict these values from more fundamental soil properties, K and D must be measured experimentally. In principle, K and D can be obtained from either steady-state or transient-state flow systems. In steady flow systems, flux, gradient, and water content are constant with time, while in transient flow systems, they vary. In general, therefore, measurements based on steady flow are more convenient to carry out and often more accurate. The difficulty, however, lies in setting up the flow system.

Techniques for measurement of conductivity and diffusivity of soil samples or models in the laboratory were described by Klute (1965). The conductivity is usually measured by applying a constant hydraulic-head difference across the sample and measuring the resulting steady flux of water. Soil samples can be desaturated either by tension-plate devices or in a pressure chamber. Measurements are made at successive levels of suction and wetness, so as to obtain the functions $K(\psi)$, $K(\theta)$, and $D(\theta)$. The $K(\psi)$ relationship is hysteretic, and therefore, to completely describe it, measurements should be made both in desorption and in sorption, as well, perhaps, as in intermediate scanning. This is difficult, however, and requires specialized apparatus (Tanner and Elrick, 1958), so that all too often only the desorption curve is measured (starting at saturation and proceeding to increase the suction in increments).

Such laboratory techniques can also be applied to the measurement of undisturbed soil cores taken from the field. This is certainly preferable to measurements taken on fragmented and artificially packed samples, though it should be understood that no field sampling technique yet available provides truly undisturbed samples.

A widely used transient-flow method for measurement of conductivity and diffusivity in the laboratory is the "outflow method" (Gardner, 1956).

It is based on measuring the falling rate of outflow from a sample in a pressure cell when the pressure is increased by a certain increment. One problem encountered in the application of this method is that of the hydraulic resistance (also called "impedance") of the plate (or membrane) and of the soil-to-plate contact zone. Techniques to account for this resistance were proposed by Miller and Elrick (1958), Rijtema (1959), and Kunze and Kirkham (1962).

Laboratory measurements of conductivity and diffusivity can also be made on long columns of soil, not only on small samples contained in cells. In such a column, steady-state flow can be induced (e.g., Moore, 1939; Youngs, 1964). If the column is long enough to allow the measurement of suction gradients (as by a series of tensiometers) and of water-content gradients (as by sectioning, or, preferably, by some nondestructive technique such as gamma-ray scanning), the $K(\theta)$ and $K(\psi)$ relationship can be obtained for a considerable range of θ with a single column or with a series of columns.

Measurements in columns under transient flow conditions have also been made (e.g., the horizontal column technique of Bruce and Klute, 1956). If periodic suction and wetness profiles are measured, the flux values at different time and space intervals can be evaluated by graphic integration between successive moisture profiles. This procedure has been called the "instantaneous profile" technique, and it can be applied in the field as well (Watson, 1966; Rose et al., 1965).

It is generally more difficult to set up steady-flow regimes in the field than in the laboratory. Infiltration techniques have been proposed by Youngs (1964) based on a steady application rate by sprinkling, and by Hillel and Gardner (1970) based on infiltration through a series of capping crusts. The effect of an impeding layer (crust) present at the boundary of entry during infiltration is to decrease the potential at the soil surface, thus reducing the driving force, and to decrease the soil water content (and correspondingly, the conductivity and diffusivity values) of the infiltrating column.

An additional field method for measuring diffusivity in an internally draining profile was recently proposed by Gardner (1970). In this procedure, the profile is wetted deeply and allowed to drain while evaporation from the soil surface is prevented. The experiment must be done in the absence of water uptake by plants or other sources or sinks.

Integration of the unsaturated-flow equation in one dimension once with respect to depth gives

$$\int_0^L \frac{\partial \theta}{\partial t}\, dz = K \frac{\partial H}{\partial z} \tag{5.22}$$

where θ is volumetric wetness, t time, z depth, K unsaturated hydraulic conductivity, and H hydraulic head. The left side of Eq. (5.22) represents the rate of water loss from that portion of the profile, which can be determined in the field from successive water content measurements. If one knows $\partial H/\partial z$, it is a simple matter to calculate K. As the profile drains, K can be obtained as a function of the soil water content.

In a homogeneous profile in the absence of a shallow water table, the hydraulic-head gradient is often very nearly unity, so that the drainage rate is almost exactly the conductivity (Black *et al.*, 1969). Where the profile drains fairly uniformly, θ can be assumed to be a function of time but not of depth, and Eq. (5.22) reduces to

$$L\frac{\partial \bar{\theta}}{\partial t} = -K\frac{\partial H}{\partial z}\bigg|_L \qquad (5.23)$$

where $\bar{\theta}$ is the average wetness above depth L, and the conductivity and hydraulic gradient are evaluated at the depth L. If we assume a unique relation between $\bar{\theta}$ and matric suction ψ, we can write

$$L\frac{d\theta}{d\psi}\frac{\partial \psi}{\partial t} = -K\frac{\partial H}{\partial z}\bigg|_L \qquad (5.24)$$

Rearranging terms and remembering that by definition the diffusivity $D = -K(d\psi/d\theta)$, we have

$$D = L\frac{d\psi/dt}{dH/dz} \qquad (5.25)$$

Thus, D can be determined from the time rate of change of the matric suction, and the hydraulic gradient. In cases where the hydraulic gradient is nearly unity, only the time rate of change of the matric suction is needed. The only instrumentation required is one tensiometer, or, preferably, several tensiometers at various depths in the profile. If the soil is not draining uniformly, the diffusivity calculated will be an average over the entire profile above the depth L.

H. Vapor Movement

We have already stated that liquid water moves in the soil by *mass flow*, a process by which the entire body of a fluid flows in response to differences in total pressure. In certain special circumstances, water-vapor movement can also occur as mass flow; for instance, when wind gusts induce bulk movement of air and vapor mixing in the surface zone of the soil. In general, however, vapor movement through most of the soil profile occurs by *diffusion*, a process in which different components of a mixed fluid move independently, and at times in opposite directions, in response to differences in concentration (or partial pressure) from one location to another. Water vapor is always present in the gaseous phase of an unsaturated soil, and vapor diffusion occurs whenever differences in vapor pressure develop within the soil.

The diffusion equation for water vapor is

$$q_d = -D_{vap} \frac{\Delta P_{vap}}{L} \tag{5.26}$$

where q_d is the diffusion flux, D_{vap} the diffusion coefficient for water vapor, ΔP_{vap} the vapor-pressure difference between two points in the soil a distance L apart (i.e., $\Delta P/L$ is the vapor-pressure gradient). D_{vap} in the soil is lower than in open air, because of restricted porosity and the tortuosity of the air-filled pores (Currie, 1961).

At constant temperature, the vapor-pressure differences which may develop in a nonsaline soil are likely to be very small. For example, a change in matric suction between 0 and 100 bars is accompanied by a vapor pressure change of only 17.54 to 16.34 Torr, a difference of only 1.6 millibar. For this reason, it is generally assumed that under normal field conditions soil air is nearly vapor-saturated at almost all times. Vapor-pressure gradients can be caused by differences in the concentration of dissolved salts, but this effect is probably appreciable only in saline soils.

When temperature differences occur, however, they might cause considerable differences in vapor pressure. For example, a change in water temperature from 19° to 20°C results in an increase in vapor pressure of 1.1 Torr. In other words, a change in temperature of 1°C has nearly the same effect upon vapor pressure as a change in suction of 100 bars!

In the range of temperatures prevailing in the field, the variation of saturated vapor pressure (that is, the vapor pressure in equilibrium with pure, free water) is as follows:

Temperature °C	0	20	30	40
Vapor pressure (Torr)	4.58	17.5	38.0	55.8

Vapor movement tends to take place from warm to cold parts of the soil. Since during the daytime the soil surface is warmer, and during the night colder, than the deeper layers, vapor movement tends to be downward during the day and upward during the night. Temperature gradients can also induce liquid flow.

Since liquid movement includes the solutes, while vapor flow does not, there have been attempts to separate the two mechanisms by monitoring salt movement in the soil (Gurr et al., 1952; Deryaguin and Melnikova, 1958). It has been observed that the rate of vapor movement often exceeds the rate which could be predicted on the basis of diffusion alone (Cary and Taylor, 1962). It appears to be impossible to separate absolutely the liquid from the vapor movement, as overall flow can consist of a complex sequential process of evaporation, short-range diffusion, condensation in capillary pockets of liquid, short-range liquid flow, reevaporation, etc. (Philip and de Vries, 1957). The two phases apparently move simultaneously and interdependently as a consequence of the suction and vapor-pressure gradients in the soil. It is commonly assumed, however, that liquid flow is the dominant mode in moist, nearly isothermal soils (Miller and Klute, 1967), and hence that the contribution of vapor diffusion to overall water movement is negligible in the main part of the root zone where diurnal temperature fluctuations are slight.

I. Water Transport Associated with Thermal Gradients

The fact that temperature gradients can induce water movement in soils has been known for at least 50 years (Bouyoucos, 1915).

Studies on the relative importance and the interaction of thermal and suction gradients in transporting soil moisture were carried out by Hutchison, et al. (1948), Philip and de Vries (1957), Taylor and Cary (1960), Cary and Taylor (1962) and Cary (1965, 1966). In analyses of the simultaneous transport of water and heat, the equations of irreversible thermodynamics, and particularly the Onsager reciprocity relations,[10] are used (Cary, 1963).

[10] As pointed out in Chapter 3, classical thermodynamics deals with reversible processes and equilibrium states. It can predict whether, and in what direction (but not at what rate) a spontaneous process will occur in a system not at equilibrium. In natural systems, any number of different forces X_i might be operating simultaneously to produce mutually interacting fluxes, J_i (e.g., a concentration gradient causes diffusion, while a pressure gradient induces mass flow and a temperature gradient results in the transfer of heat, with each of these fluxes involving the others on the same system). If the system is not too far from equilibrium, the fluxes are taken to be related linearly to the forces causing them. Thus:

$$J_1 = L_{11}X_1 + L_{12}X_2 - \ldots + L_{1n}X_n$$
$$J_2 = L_{21}X_1 + L_{22}X_2 - \ldots + L_{2n}X_n$$
.
.
.
$$J_n = L_{n1}X_1 + L_{n2}X_2 - \ldots + L_{nn}X_n$$

or:

$$J_i = \sum_{k=1}^{n} L_{ik}X_k$$

where L_{ik} are the transmission coefficients of the various fluxes (e.g., the diffusion coefficient, the hydraulic conductivity, the thermal conductivity, etc.). According to Onsager's principle of reciprocity, the phenomenological cross-coefficients are equal ($L_{ik} = L_{ik}$). The validity of these relations may fail if the systems is very far from equilibrium since the fluxes may no longer be linear with the forces. For a more complete analysis of this topic the reader is referred to the treatises by Prigogine (1961) and de Groot (1963).

Philip and de Vries (1957) presented a general differential equation describing moisture movement in porous materials under combined temperature and moisture gradients for one-dimensional vertical flow as follows:

$$\frac{\partial \theta}{\partial t} = \nabla \cdot (D_T \nabla T) + \nabla \cdot (D_w \nabla \theta) - \frac{\partial K}{\partial z} \qquad (5.27)$$

where θ is moisture content, t time, T absolute temperature, D_T thermal diffusivity $[D_T(\text{liq}) + D_T(\text{vap})]$, K hydraulic conductivity of the soil, z vertical space coordinate, and D_w moisture diffusivity $[D_w(\text{liq}) + D_w(\text{vap})]$. The heat conduction equation for soil is given similarly as

$$C_v \frac{\partial T}{\partial x} = \nabla \cdot (C_{Ta} \nabla T) - H_L \nabla \cdot (D_{w,vap} \nabla \theta) \qquad (5.28)$$

where C_v is volumetric heat capacity of the soil (cal cm^{-3} °C^{-1}), C_{Ta} apparent thermal conductivity of the soil (cal sec^{-1} cm^{-1} °C^{-1}), and H_L latent heat of vaporization of water. Both the above equations are of the diffusion type involving θ- and T-dependent diffusivities as well as gradients of both θ and T. Taken together, these equations describe the simultaneous transfer of moisture and heat in soils.

Cary (1963) has developed the equations

$$J_{H_2O} = -L_{H_2O}C_p \ln \frac{T^h}{T^c} + (\mathscr{L} + ML_{H_2O})\left(\frac{1}{T^c} - \frac{1}{T^h}\right)$$

$$J_{Heat} = (L_{Heat} + \mathscr{L}M)\left(\frac{1}{T^c} - \frac{1}{T^h}\right) - \mathscr{L}C_p \ln \frac{T^h}{T^c} \qquad (5.29)$$

to apply to water movement in soil under a thermal gradient. In these equations, C_p is the heat capacity of water; T^h and T^c are respectively the temperatures of the hot and cold ends of the soil column; M is an integration constant; $\mathscr{L}$ is the coefficient of interaction between heat and moisture flow; J_{H_2O} and J_{Heat} are the fluxes; and L_{H_2O} and L_{Heat} are the transmission coefficients for water and heat, respectively.

J. Movement of Solutes

Soil water is invariably a solution, containing various soluble substances which move both within and with the water phase. Some of the solutes may enter into or come out of the soil's exchange complex (i.e., the adsorbed phase), some may precipitate out of solution or redissolve in it, some may volatilize and escape into the atmosphere, some may be produced or consumed by biological activity (e.g., nutrient uptake by plants, mineralization of organic matter by microbes, etc.), and finally, some may be leached away during periods of drainage or percolation.

Such processes, which may be bidirectional, can determine the amounts of nutrients present and available for plant growth, as well as the salinity level and status of the exchange complex of the soil. The soil matrix itself, and particularly the clay fraction, is sensitive to changes in the composition and concentration of the permeating solution. Hence, the study of soil-water relationships must necessarily take into consideration the state and movement of soluble substances.

In this book, we have chosen, in the interest of brevity, to concentrate upon purely physical relationships and to disregard, *as far as possible*, complications due to chemical processes involving the soil solution. Yet it is not possible to disregard these entirely, and at least an elementary elucidation of phenomena involving solutes is necessary. In including this section in the chapter on unsaturated flow we do not wish to imply that solute movement does not occur under saturated conditions as well. Indeed, it does. However, since unsaturated conditions prevail generally in the rooting zones of most plant habitats of interest (particularly in agricultural fields), it is only proper to emphasize the aspects of solute movement characteristic of unsaturated soils.

Comprehensive reviews of the interactions of solutes in soils were published by Gardner (1965) and Bresler (1970). The following discussion is based in part on these reviews.

Substances dissolved in the soil solution can move by molecular or ionic *diffusion*, due to concentration gradients within the solution; or by *convection*, due to the mass flow of the soil solution. The processes of diffusion and convection can occur simultaneously, either in the same direction or in opposition. Movement of ions can also be affected by electric fields of clay surfaces. Thus, a strongly adsorbed cation will move less readily than weakly adsorbed cations or anions.

In treating diffusion in soils we must take into account the fact that diffusion can only occur in that fraction of the soil which is filled with water (or air, in the case of vapor diffusion); thus Fick's first law may be written for a one-dimensional system

$$q_d = -D\theta \frac{dC}{dx} \tag{5.30}$$

where q_d is the flux of the diffusing substance per unit area of soil, D the effective diffusion coefficient, θ the volume fraction of water (wetness), C the concentration of the diffusing substance in soil water, and x the space variable. Because of tortuosity and the effect of adsorbed water, D in the soil is generally less than in bulk water. In the case of a saturated soil, θ is equal to the porosity.

The diffusion coefficient decreases with decreasing wetness θ, so that q_d is strongly dependent upon the degree of unsaturation. This is borne out by the data of Porter *et al.* (1960) and Romkens and Bruce (1964).

D also depends upon temperature and to some extent upon concentration. To preserve electrical neutrality, the movement of any ion must be accompanied by the diffusion of some other ion or ions. The effective diffusion coefficient for a combination of ions is some average of the separate coefficients for the different ionic species (Low, 1962).

To describe nonsteady diffusion, the continuity equation is applied to Eq. (5.30), resulting in a second-order equation analogous to Fick's second law of diffusion:

$$\frac{\partial M}{\partial t} = D\theta \frac{\partial^2 C}{\partial x^2} \tag{5.31}$$

or

$$\frac{\partial C}{\partial t} = D \frac{\partial^2 C}{\partial x^2} \tag{5.32}$$

where M is the quantity of substance per unit volume of soil and t is time. In applying Eqs. (5.31) and (5.32), isothermal conditions are usually assumed.

If the soil solution itself is flowing within the soil, then obviously the solutes are carried along. Neglecting "salt sieving" and other interaction effects (Kemper, 1960), the flux of solutes q_s due to water movement should be

$$q_s = q_w C$$

where q_w is the flux of water. This flux can be related to the average velocity of water in the soil $\bar{u}$ by the expression $q_w = \bar{u}\theta$. The rate of change of solute content per unit volume of soil due to convection is

$$\frac{\partial M}{\partial t} = -\frac{\partial q_s}{\partial x} = -\frac{\partial (q_w C)}{\partial x} \tag{5.33}$$

when diffusion and convection take place simultaneously, Eqs. (5.32) and (5.33) can be combined to give

$$\frac{\partial M}{\partial t} = D \frac{\partial^2 C}{\partial x^2} - \theta \bar{u} \frac{\partial C}{\partial x}$$

$$\frac{\partial C}{\partial t} = D \frac{\partial^2 C}{\partial x^2} - \frac{\partial (\bar{u} C)}{\partial x} \tag{5.34}$$

The process of diffusion plus convection is more complicated than is indicated by Eq. (5.34). Since the actual velocity of water flowing in pores is not uniform (e.g., the velocity near the center of a pore's cross section exceeds that near the edge, and the velocity in wide pores exeeds that in narrow or constricted or lateral pores), the flow results in a mixing process known as *hydrodynamic dispersion*. Thus, the effective diffusion coefficient of Eq. (5.34) depends upon flow velocity, and tends to increase with increasing flux (Nielsen and Biggar, 1963). Work on hydrodynamic dispersion in soil was reported by Day (1956) and Day and Forsythe (1957). An extensive review and analysis of hydrodynamic dispersion is included in the recent book by Bear, *et al.* (1968).[11]

The movement of exchangeable ions in the soil is difficult to describe quantitatively, particularly in view of the fact that exchange reactions are reversible and hence the presence and concentration of all competing ions must be taken into account. The mass concentration of a given ionic species in the soil now consists of its concentration in the solution plus its concentration in the adsorbed phase, the latter being a function (usually nonlinear) of the former. Work on this problem was reported by Bower et al. (1957).

In the special case in which the rate of adsorption is linear with concentration, Eq. (5.34) can be rewritten as:

$$\frac{\partial C}{\partial t} = D \frac{\partial^2 C}{\partial x^2} - \frac{\partial(\bar{u}C)}{\partial x} + \lambda C \qquad (5.35)$$

where λ is the exchange rate coefficient.

An important practical aspect of solute movement is the *leaching* process, which is essential to salinity control in irrigation. When irrigation is practiced in arid regions, particularly if the irrigation water contains an appreciable concentration of soluble salts, it is necessary to allocate a certain amount of water beyond the transpirational requirement in order to ensure the removal of excess soluble salts from the root zone. Without leaching, salt will accumulate in direct proportion to the salt content of the irrigation water and the depth of water applied, since plants generally exclude nearly all of the salt and hence the process of evapotranspiration concentrates the soil solution.

It has been shown (Richards, 1954) that less than 2 ft of reasonably good-quality irrigation water contains sufficient salt to salinize an initially saltfree loam soil, if there is no leaching or precipitation of salt in the soil. For leaching to be effective, the soil must obviously be well drained. In some areas where natural drainage is slow and artificial drainage cannot be provided, it becomes impossible to sustain irrigation and the land must sooner or later be abandoned owing to salinization.

---

[11] The dispersion phenomenon is illustrated in the shape of so-called *breakthrough curves*. If a slug of tracer-marked fluid is applied at a certain moment to the inflow face of a conducting porous column and the tracer concentration at the end of the column is measured and plotted against time, then it can be observed that the abrupt concentration change applied at the inflow generally degenerates into an S-shaped curve at the end of the column. Such breakthrough curves for soils were reported by Nielsen and Biggar (1963).

The *leaching requirement* concept (Richards, 1954) is based on the use of the excess infiltration over evapotranspiration to estimate the amount of leaching. It has been defined as the fraction of the irrigation water that must be leached through the root zone to control soil salinity at any specific level. The leaching requirement depends upon the salt concentration of the irrigation water and upon the maximum concentration permissible in the soil solution.[12] It also depends upon the amount of water extracted from the soil by the roots and upon the salt tolerance of the crop. An equation given by Richards is

$$d_i = \left(\frac{E_d}{E_d - E_i}\right) d_{et} \tag{5.36}$$

where d_i is the depth of irrigation, d_{et} the equivalent depth of consumptive use by the plants (mainly by evapotranspiration), and E_d and E_i are the electrical conductivities of the drainage water and of the irrigation water, respectively. Theoretically based equations were later attempted by van der Molen (1956), who used the theory of chromatography and found agreement between theoretical curves and measured leaching rates in Dutch soils which had been inundated by seawater.

Gardner and Brooks (1956) developed a theory in which a distinction was made between immobile (detained) and mobile salt, the latter moving with the velocity of the leaching front. They observed that in several soils about 1.42 pore volumes of water are needed to reduce salinity by 80%, and held that the flow process itself, rather than diffusion, is responsible both for the diffuse boundary between the soil solution and the leaching water and for the subsequent removal of the temporarily bypassed salt.

Nielsen and Biggar (1961) conducted studies on miscible displacement and applied these to the leaching of excess salts from saline soils. They suggested that leaching soils at a water content below saturation (e.g., under sprinkling irrigation or rainfall or under intermittent irrigation) could produce more efficient leaching and thereby reduce the amount of water required as well as reduce drainage problems in areas of high water tables. This corroborated observations that in a soil with large vertical cracks most of the water infiltrated under ponding moves through these cracks and is ineffective in leaching, whereas under rainfall, more of the water moves through the soil blocks and micropores, producing more efficient leaching.

---

[12] According to the standards of the U.S. Salinity Laboratory, the maximum concentration of the soil solution should be kept below 4 mmhos/cm (expressed in terms of electrical conductivity) for sensitive crops. Tolerant crops like beets, alfalfa, and cotton may give good yields at values up to 8 mmhos/cm, while a very tolerant crop like barley may give good yields at 12 mmhos/cm.

The movement and distribution of salts in the soil profile during and after successive irrigations were studied by Bresler (1967) and by Bresler and Hanks (1969), who used numerical methods of analysis. Nutrient movement to plant roots was studied by Bray (1954), Barber (1962) and Olsen and Kemper (1968).

K. Soil Aeration

The process of soil aeration is one of the most important determinants of soil productivity. Plants absorb oxygen and release carbon dioxide in the process of respiration. In most terrestrial plants (excepting such specialized plants as rice), the internal transfer of oxygen from the above-ground parts to the roots cannot take place at a rate sufficient to supply the oxygen requirements of the roots. Adequate root growth requires that the soil be *aerated*, that is to say, that gaseous exchange take place between soil air and the atmosphere at such a rate as to prevent a deficiency of oxygen and an excess of carbon dioxide from developing in the root zone. Soil microorganisms also respire, and, under conditions of restricted aeration, might compete with the roots of higher plants (Stotzky, 1965).

Gases can move either in the air phase (that is, in the pores which are drained of water, provided they are interconnected and open to the atmosphere), or in dissolved form through the water phase. The rate of diffusion of gases in the air phase is generally greater than in the water phase, hence soil aeration is dependent largely upon the volume fraction of air-filled pores.

The subject of soil aeration was reviewed by Russell (1952), Erickson and van Doren (1960), Stolzy and Letey (1964), and Vilain (1963).

Measurements of oxygen-consumption rates by plant roots and soil microorganisms (Hawkins, 1962) have shown that soil aeration in the field can involve the exchange of approximately 10 liters of oxygen per square meter per day. Assuming that the air-filled porosity is 20% of the soil volume, that soil air contains 20% oxygen, and that the root zone is 1 m deep, then the total root-zone content of oxygen is 40 liters and hence the daily exchange must involve 25% of the oxygen present. However, these figures are given only to provide an order of magnitude, since in actual conditions the aeration rate probably varies between wide limits. Plant growth probably depends more upon the occurrence and duration of periods of oxygen deficiency than upon average conditions (Erickson and van Doren, 1964).

Gaseous exchange between soil and atmosphere can occur as mass flow or as diffusion. Mass flow can result from the penetration of wind or convection currents into large soil pores (or cavities), or from temperature or barometric pressure changes as well as from water entry and withdrawal. It is generally recognized, however, that the principal mechanism in the interchange of soil air is the diffusion process, rather than mass flow (Evans, 1965).

The rate of diffusion of gases in the soil increases with the air-filled porosity and depends only secondarily on the size distribution of these pores (Penman, 1940; Marshall, 1959).[13] Soil aeration is likely to become limiting to plant growth when the air-filled porosity falls below about 10% (Baver and Farnsworth, 1940; Vomocil and Flocker, 1961), though it must be realized that it is the rate of exchange rather than simply the content of soil air that is the decisive factor.

The composition of soil air (van Bavel, 1965) depends on aeration conditions. In a well-aerated soil, the air phase does not differ significantly from the composition of the atmosphere, except for a higher relative humidity (soil air is generally nearly vapor-saturated) and higher concentration of carbon dioxide (i.e., 0.2–1% in soil air as compared to 0.03% in the atmosphere. In a soil with restricted aeration (e.g., under flooding), the oxygen concentration can decrease and carbon dioxide concentration can increase very markedly. If aeration is restricted for long, chemical reduction takes place and such gases as methane, nitrous oxide, and hydrogen sulfide might evolve.

Soil aeration and soil-water relations interact not only because the relative amount of water present determines the air-filled porosity and diffusion rate, but also because air dissolves in and diffuses through soil water. Oxygen is only 4% as soluble as carbon dioxide, and its diffusion coefficient in water is of the order of 0.01% of its diffusion coefficient in air. The movement of oxygen to roots can be limited by the rate of diffusion through the water films enveloping plant roots.

Poor aeration can decrease the uptake of water and induce early wilting (Stolzy et al., 1963). According to Kramer (1956), restricted aeration causes a decrease in the permeability of roots to water.

L. Summary

Most of the processes of water flow in agricultural soils occur under unsaturated conditions. Darcy's law has been found to apply for unsaturated as well as for saturated soils, but the pressure gradient at unsaturation becomes a suction gradient, and the hydraulic conductivity is no longer

---

[13] According to Penman (1940), the apparent diffusion coefficient for gases in the air phase of the soil is about 2/3 the diffusion coefficient in bulk air, owing to the tortuosity of soil pores.

constant, but a function of water content or suction. Since the conductivity depends on the number, sizes, and shapes of the conducting pores, its value is greatest when the soil is saturated, and decreases steeply when the soil-water suction increases and the soil loses moisture. Darcy's law suffices to describe water flow under steady-state conditions, but must be combined with the continuity equation to describe unsteady (transient-state) flow. In the water-content and suction ranges prevailing throughout most of the root zone in the field, liquid movement predominates, but vapor movement, and particularly the combined movement of vapor and liquid, can be important where appreciable temperature gradients occur. Liquid water movement entails the movement of solutes, the concentration and accumulation of of which must be controlled (e.g., to prevent soil salinization in arid regions, or excessive leaching of nutrients in humid regions. Unsaturated soils contain an air phase, which, if open to the atmosphere, can sustain the respiration requirements of plant roots. Restricted soil aeration (as in water-logged soils) can hinder plant growth.

Part II: *THE FIELD WATER CYCLE*

The cyclic movement of water in the field begins with its entry into the soil by the process of infiltration, continues with its temporary storage in the rooting zone, and ends with its removal from the soil by drainage, evaporation, or plant uptake. This cycle consists of a number of fairly distinct stages, or processes, which may occur simultaneously and interdependently, but which, for the sake of clarity, we shall attempt to describe separately in the following chapters.

6 Infiltration—Entry of Water into Soil

A. General

Infiltration is the term applied to the process of water entry into the soil, generally (but not necessarily[1]) through the soil surface and vertically downward. This process is of great practical importance, since its rate often determines the amount of runoff which will form over the soil surface (and hence also determines the hazard of erosion) during rainstorms. Where the rate of infiltration is limiting, the entire water economy of the rooting zone of plants may be affected. Knowledge of the infiltration process as it relates to soil properties and mode of water supply is needed for efficient soil and water management. Comprehensive reviews of infiltration processes were published by Parr and Bertrand (1960) and by Philip (1969).

B. Description of the Process

If we sprinkle or otherwise supply water to the soil surface at a steadily increasing rate, sooner or later the moment will come when the supply rate begins to exceed the capability of the soil to absorb water, and the excess will accumulate and pond over the surface, or trickle downslope as runoff (Fig. 6.1). The *infiltration rate* is the flux passing through the surface and

---

[1] Water may enter the soil through the entire surface uniformly, as under ponding or rain, or it may enter the soil through furrows or crevices, or it may move up into the soil from a source below (e.g., a high water table).

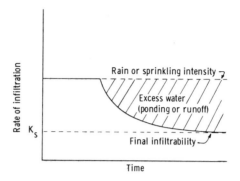

Fig. 6.1. Dependence of the infiltration rate upon time, under an irrigation of constant intensity lower than the initial value, but higher than the final value of soil infiltrability.

flowing into the profile. The *infiltration capacity*,[2] which we propose to call *soil infiltrability*, is the flux which the soil profile can absorb through its surface when it is maintained in contact with water at atmospheric pressure. As long as the rate of water supply to the surface is smaller than soil infiltrability, water infiltrates as fast as it is supplied and the supply rate determines the infiltration rate (i.e., the process is *flux-controlled*). However, once the supply rate exceeds soil infiltrability, it is the latter which determines the actual infiltration rate, and thus the process becomes *profile-controlled.*

If a shallow layer of water is instantaneously applied, and thereafter maintained, over the surface of an initially unsaturated soil, the full measure of soil infiltrability comes into play from the start. Many measurements of infiltration under shallow ponding have shown infiltrability to vary, and generally to decrease, in time. Thus, the *cumulative infiltration*, being the time integral of the infiltration rate, has a curvilinear time dependence, with a gradually decreasing slope.

Soil infiltrability and its variation with time are known to depend upon the initial wetness and suction, as well as on the texture, structure, and uniformity (or layering sequence) of the profile. In general, soil infiltrability is high in the early stages of infiltration, particularly where the soil is initially quite dry, but tends to decrease monotonically and eventually to approach asymptotically a constant rate which is often termed the *final infiltration capacity*[3] but which we prefer to call the *steady-state infiltrability.*

---

[2] The word "capacity" is generally used to denote an amount, or a volume. Its use in connection with the time-rate of a process can be misleading.

[3] The adjective "final" in this context does not signify the end of the process (since infiltration can persist practically indefinitely if profile conditions permit) but it does indicate that soil infiltrability has finally attained a constant value from which it appears to decrease no more.

Numerous empirical expressions have been proposed to describe the dependence of infiltrability upon time and upon cumulative infiltration. Among these are the equations of Kostiakov (1932) and of Horton (1940). However, purely empirical equations, not based on basic physical relationships, cannot be expected to apply universally.

The decrease of infiltrability from an initially high rate can in some cases result (at least in part) from gradual deterioration of soil structure and the consequent partial sealing of the profile by the formation of a dense surface crust, or from the detachment and migration of pore-blocking particles, or from swelling of clay, or from entrapment of air bubbles or the bulk compression of soil air if it is prevented from escaping as it is displaced by incoming water. Primarily, however, the decrease in infiltration rate results from the inevitable decrease in the matric suction gradient (constituting one of the forces drawing water into the soil) which occurs as infiltration proceeds.

If the surface of an initially dry soil is suddenly saturated, the matric suction gradient acting in the surface layer is at first very steep. As the wetted zone deepens, however, this gradient is reduced, and, as the wetted part of the profile becomes thicker and thicker, the suction gradient tends eventually to become vanishingly small. In a horizontal column, the infiltration rate eventually tends to zero, whereas in downward flow into a vertical column, the infiltration rate can be expected to settle down to a steady rate, gravity-induced, which, as we shall later show, is practically equal to the saturated hydraulic conductivity if the profile is homogeneous and structurally stable. If the surface is supplied with water at a rate lower than the saturated conductivity, or is otherwise maintained at a wetness lower than saturation, then the steady infiltration rate will correspond to the unsaturated conductivity at the particular wetness obtained.

C. Profile Moisture Distribution during Infiltration

If we examine a homogeneous profile at any moment during infiltration under ponding, we shall find that the surface of the soil is saturated, perhaps to a depth of several millimeters or centimeters, and that beneath this zone of complete saturation is a lengthening zone of apparently uniform, nearly saturated, wetness, known as the *transmission zone*. Beyond this zone there is a *wetting zone*, in which soil wetness decreases with depth at a steepening gradient down to a *wetting front*, where the moisture gradient is so steep[4]

---

[4] The reason for the steepening gradient is that, as the water content decreases, the hydraulic conductivity generally decreases exponentially. Since the flux is the product of the gradient by the conductivity, it follows that, to get a certain flux moving in the soil, the gradient must increase as the conductivity decreases.

that there appears to be a sharp boundary between the moistened soil above and the dry soil beneath.

The typical moisture profile during infiltration, first described by Bodman and Coleman (1944) and Coleman and Bodman (1945), is shown schematically in Fig. 6.2. Later investigations have cast some doubt as to whether a saturation zone distinct from the transmission zone necessarily exists, or whether it is merely an experimental artifact or anomaly resulting from the looseness, structural instability, slaking, or swelling of the soil at the surface. The saturation zone might also result from the air-entry value, or from air

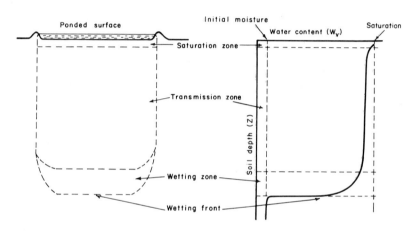

Fig. 6.2. The infiltration moisture profile. At left, a schematic section of the profile; at right, the water-content vs. depth curve.

entrapment. The surface soil, being unconfined and subject to the disruptive and slaking action of raindrops and turbulent water, often experiences aggregate breakdown and colloidal dispersion, resulting in the formation of an impeding crust, which, in turn, affects the moisture profile below.

If we continue to examine the moisture profile periodically during infiltration, we shall find that the nearly saturated transmission zone lengthens (deepens) continuously, and that the wetting zone and wetting front move downward continuously, with the latter becoming less steep as it moves deeper into the profile. Typical families of successive moisture and hydraulic-head profiles are shown in Fig. 6.3.

With the foregoing qualitative description of infiltration as a background, we can now proceed to consider some of the quantitative aspects of the process as it occurs under various conditions.

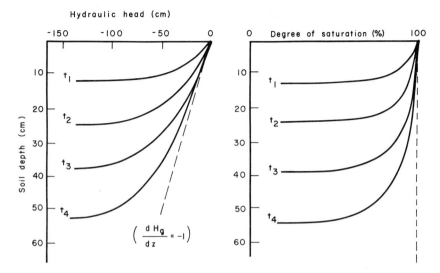

Fig. 6.3. Water-content profiles (at right) and hydraulic-head profiles (at left) at different times ($t_1 < t_2 < t_3 < t_4$) during infiltration into a uniform soil ponded at the surface. dH_g/dZ is the gravitational head gradient.

D. Horizontal Infiltration

The simplest case of infiltration is that in which gravity is zero or negligible, and water is drawn into the soil by matric suction forces only. In this case (studied by Philip, 1955; Bruce and Klute, 1956; Gardner and Mayhugh, 1958; Nielsen *et al.*, 1962; Ashcroft *et al.*, 1962; and Ferguson and Gardner, 1963), the diffusivity equation can be applied directly (subject to the limitations described in Chapter 5):

$$\frac{\partial \theta}{\partial t} = \frac{\partial}{\partial x}\left[D(\theta)\frac{\partial \theta}{\partial x}\right] \tag{6.1}$$

This equation can be solved readily for the special conditions of flow into an infinitely long column of uniform initial wetness θ_i, where the plane of entry ($x = 0$) is instantaneously brought to, and thereafter maintained at, a higher wetness θ_0. These conditions are written formally in the following way:

$$\theta = \theta_i, \qquad x \geq 0, \quad t = 0$$

$$\theta = \theta_0, \qquad x = 0, \quad t > 0 \tag{6.2}$$

For certain known functions of $D(\theta)$, a solution is obtainable in closed form for the ordinary differential equation resulting from the Boltzmann transformation, $y = (x/2)\sqrt{D_i t}$ (Gardner and Mayhugh, 1958). The solution indicates that the rate of advance of any wetness value, and the infiltration rate itself, are reciprocally proportional to the square root of time, while the distance of advance of any wetness value, and the cumulative infiltration, are directly proportional to the square root of time. For the entry of water into soils, exponential D functions of the form

$$D = D_i \exp(\theta - \theta_i) \tag{6.3}$$

(where D_i is the diffusivity corresponding to the initial wetness θ_i) were found to give good agreement between theory and experiment for a number of soils. Other functions exist which give solutions but few actually apply to soils.

The time dependences of the cumulative infiltration I and of the infiltration rate i can be expressed in terms of the *weighted mean diffusivity*[5] $\bar{D}$ (Crank, 1956), which is a constant value of the diffusivity giving the same flux as the variable diffusivity which in fact operates during the actual flow process. Assuming the diffusivity constant allows equation 6.1 to be solved analytically. The following equations are obtained:

$$i = \frac{1}{2}(\theta_0 - \theta_i)\sqrt{\frac{\bar{D}}{\pi t}} \tag{6.4}$$

and

$$I = \int_0^t i \, dt = (\theta_0 - \theta_i)\sqrt{\frac{\bar{D}t}{\pi}} \tag{6.5}$$

which, again, indicate square-root-of-time behavior. Time t can be eliminated from these equations to obtain the relation of infiltration rate to cumulative infiltration:

$$i = (\theta_0 - \theta_i)^2 \bar{D}/\pi I \tag{6.6}$$

As mentioned earlier, when water moves into a relatively dry soil, a distinct wetting front is often observed, this front being in effect a moving boundary between the already wetted and the as yet unwetted parts of the soil. From the considerations given, it can be inferred that the steepness, or

---

[5] According to Crank (1956), the weighted mean diffusivity for sorption processes is given by

$$\bar{D} = (5/3)[1/(\theta_0 - \theta_i)]^{5/3} \int_{\theta_1}^{\theta_0} D(\theta)(\theta - \theta_i)^{2/3} \, d\theta$$

sharpness, of the wetting front is related to the difference between the diffusivity of the wetted soil (near the entry surface) and that of the relatively dry soil ahead of the wetting front. Thus, a coarse-textured soil, which characteristically shows steeper decrease of D with decreasing θ, typically exhibits a sharper and more distinct wetting front than fine-textured soils. Similarly, the wetting front is sharper during infiltration into dry than into moist soils.

E. Vertical Infiltration

Downward infiltration into an initially unsaturated soil generally occurs under the combined influence of suction and gravity gradients. As the water penetrates deeper and the wetted part of the profile lengthens, the average suction gradient decreases, since the overall difference in pressure head (between the saturated soil surface and the unwetted soil inside the profile) divides itself along an ever-increasing distance. This trend continues until eventually the suction gradient in the upper part of the profile becomes negligible, leaving the constant gravitational gradient as the only remaining force moving water downward in this upper or transmission zone. Since the gravitational head gradient has the value of unity (the gravitational head decreasing at the rate of 1 cm with each centimeter of vertical depth below the surface), it follows that the flux tends to approach the hydraulic conductivity as a limiting value. In a uniform soil (without crust) under prolonged ponding, the water content of the wetted zone approaches the saturated hydraulic conductivity.

Darcy's equation for vertical flow is:

$$q = -K\frac{dH}{dz} = -K\frac{d}{dz}(H_p - z) \tag{6.7a}$$

In an unsaturated soil, H_p is negative, and can be expressed as a suction head ψ:

$$q = K\frac{d\psi}{dz} + K \tag{6.7b}$$

where q is the flux, H the total hydraulic head, H_p the pressure head, ψ the matric suction head, z the vertical distance from the soil surface downward (i.e., the depth), and K the hydraulic conductivity. At the soil surface, $q = i$, the infiltration rate.

Combined with the equation of continuity, Eq. (6.7b) becomes

$$\frac{\partial\theta}{\partial t} = -\frac{\partial}{\partial z}\left(K\frac{\partial\psi}{\partial z}\right) - \frac{\partial K}{\partial z} \tag{6.8a}$$

or

$$\frac{\partial \theta}{\partial t} = \frac{\partial}{\partial z}\left(D\,\frac{\partial \theta}{\partial z}\right) - \frac{\partial K}{\partial z} \tag{6.8b}$$

When infiltration takes place into an initially dry soil, the suction gradients are at first much greater than the gravitational gradient, and the initial infiltration rate into a vertical column is close to the infiltration rate into a horizontal column. Water from a furrow will therefore tend at first to infiltrate laterally almost to the same extent as vertically (Fig. 6.4). On the other hand,

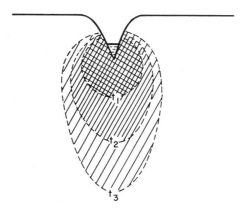

Fig. 6.4. Infiltration from an irrigation furrow into an initially dry soil. The wetting front is shown after different periods of time ($t_1 < t_2 < t_3$). At first, the strong suction gradients cause infiltration to be nearly uniform in all directions; eventually, the suction gradients decrease and the gravitational gradient predominates.

when infiltration takes place into an initially wet soil, the suction gradients are small from the start and become negligible much sooner.

Philip (1957) developed the following method of solution of Eq. (6.8), subject to the conditions

$$t = 0, \qquad z > 0, \qquad \theta = \theta_i$$
$$t \geq 0, \qquad z = 0, \qquad \theta = \theta_0 \tag{6.9}$$

His solution is of the form of a power series:

$$z(\theta,t) = \sum_{n=1}^{\infty} f_n(\theta)t^{n/2}$$
$$= f_1(\theta)t^{1/2} + f_2(\theta)t + f_3(\theta)t^{3/2} + f_4(\theta)t^2 + \cdots \tag{6.10}$$

where z is the depth to any particular value of wetness θ, and the coefficients $f_n(\theta)$ are calculated successively from the diffusivity and conductivity functions.

This solution indicates that at small times the advance of any θ value proceeds as $\sqrt{t}$ (just as in horizontal infiltration), while at larger times the downward advance of soil wetness approaches a constant rate $(K_0 - K_i)/(\theta_0 - \theta_i)$, where K_0 and K_i are the conductivities at the wetness values of θ_0 (wetted surface) and θ_i (initial soil wetness), respectively.

Philip's solution also describes the time dependence of cumulative infiltration I in terms of a power series:

$$I(t) = \sum_{n=1}^{\infty} j_n(\theta)t^{n/2}$$
$$= st^{1/2} + (A_2 + K_0)t + A_3 t^{3/2} + A_4 t^2 + \cdots \qquad (6.11)$$

in which the coefficients $j_n(\theta)$ are, again, calculated from $K(\theta)$ and $D(\theta)$, and coefficient s is called the *sorptivity*. Differentiating Eq. (6.11) with respect to t, we obtain the series for the infiltration rate $i(t)$:

$$i(t) = \tfrac{1}{2}st^{-1/2} + (A_2 + K_0) + \tfrac{3}{2}A_3 t^{1/2} + 2A_4 t + \cdots \qquad (6.12)$$

In practice, it is generally sufficient for an approximate description of infiltration to replace Eqs. (6.11) and (6.12) by two-parameter equations of the type

$$I(t) = st^{1/2} + At, \qquad i(t) = \tfrac{1}{2}st^{-1/2} + A \qquad (6.13)$$

where t is not too large. In the limit, as t approaches infinity, the infiltration rate decreases monotonically to its final asymptotic value $i(\infty)$. Philip (1969) pointed out that this does not imply that $A = K_0$, particularly not at small or intermediate times. However, at very *large times* (for which the infinite series does not converge), it is possible to represent Eq. (6.13) as

$$I = st^{1/2} + Kt, \qquad i = \tfrac{1}{2}st^{-1/2} + K \qquad (6.14)$$

where K is the hydraulic conductivity of the soil's upper layer (the transmission zone), which, in a uniform soil under ponding, is approximately equal to the saturated conductivity K_s.

The sorptivity has been defined (Philip, 1969) in terms of the horizontal infiltration equation

$$s = I/t^{1/2} \qquad (6.15)$$

As such, it embodies in a single parameter the influence of the matric suction and conductivity on the transient flow process that follows a step-function change in surface wetness or suction. Strictly speaking, one should write $s(\theta_0, \theta_i)$ or $s(\psi_0, \psi_i)$, since s has meaning only in relation to an initial state of the medium and an imposed boundary condition. The dimensions of s are $LT^{-\frac{1}{2}}$. Philip also defined an "intrinsic sorptivity"—a parameter which takes into account the viscosity and surface tension of the fluid.

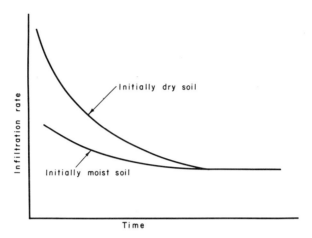

Fig. 6.5. Infiltrability as a function of time in an initially dry and in an initially moist soil.

It should be obvious from the foregoing that the effects of ponding depth and initial wetness (Fig. 6.5) can be significant during early stages of infiltration, but decrease in time and eventually tend to vanish in a very deeply wetted profile. Typical values of the "final" infiltration rate are shown in Table 6.1. These values merely give an order of magnitude, while in actual

Table 6.1

Soil type	Final infiltration rate mm/hr
Sands	>20
Sandy and silty soils	10–20
Loams	5–10
Clayey soils	1–5
Sodic clayey soils	<1

cases the infiltration rate can be considerably higher (particularly during the initial stages of the process and in well-aggregated or cracked soils), or lower (as in the presence of a crust).

F. Green and Ampt's Approach

A much simplified, approximate approach to the infiltration problem was suggested more than half a century ago by Green and Ampt (1911), in their classic paper on flow of air and water through soils. Their approach[6] has been found to apply quite satisfactorily for certain cases of infiltration into initially dry soils, particularly of the coarse-textured type, which exhibit a sharp wetting front (e.g., Hillel and Gardner, 1970). The solution gives no information about details of the moisture profile during infiltration, but does offer estimates of the infiltration rate and the cumulative infiltration functions of time, i.e., $i(t)$ and $I(t)$.

The principal assumptions of the Green and Ampt approach are that there exists a distinct and precisely definable wetting front, and that the matric suction at this wetting front remains effectively constant, regardless of time and position. Furthermore, this approach assumes that, behind the wetting front, the soil is uniformly wet and of constant conductivity. The wetting front is thus viewed as a plane separating a uniformly wetted in-filtrated zone from a totally uninfiltrated zone. In effect, this supposes the K vs. θ relation to be discontinuous, i.e., to change abruptly, at the suction value prevailing at the wetting front.

These assumptions simplify the flow equation, making it amenable to analytical solution. For horizontal infiltration, a Darcy-type equation can be applied directly:

$$i = \frac{dI}{dt} = K\frac{H_0 - H_f}{L_f} \qquad (6.16)$$

where i is the flux into the soil and through the transmission zone, I the cumulative infiltration, K the hydraulic conductivity of the transmission zone, H_0 the pressure head at the entry surface, H_f the effective pressure head at the wetting front, and L_f the distance from the surface to the wetting front (the length of the wetted zone). If the ponding depth is negligible and the surface is thus maintained at a pressure head of zero, we obtain simply

$$\frac{dI}{dt} = -K\frac{H_f}{L_f} = K\frac{\Delta H_p}{L_f} \qquad (6.17)$$

where ΔH_p is the pressure-head drop from the surface to the wetting front. This indicates that the infiltration rate varies linearly with the reciprocal of the distance to the wetting front.

Since a uniformly wetted zone is assumed to extend all the way to the wetting front, it follows that the cumulative infiltration I should be equal to the product of the wetting front depth L_f and the wetness increment $\Delta\theta = \theta_t - \theta_i$ (where θ_t is the transmission-zone wetness during infiltration and θ_i is the initial profile wetness which prevails beyond the wetting front);

[6] The Green and Ampt approach has been called the "delta-function solution" (Philip, 1969).

$$I = L_f \, \Delta\theta \tag{6.18}$$

(In the special case where θ_t is saturation and θ_i is zero, $I = fL_f$, where f is the porosity.) Therefore,

$$\frac{dI}{dt} = \Delta\theta \frac{dL_f}{dt} = K\frac{\Delta H_p}{L_f} = K\frac{\Delta\theta \cdot \Delta H_p}{I} \tag{6.19}$$

where dL_f/dt is the rate of advance of the wetting front. The infiltration rate is thus seen to be inversely related to the cumulative infiltration. Rearranging Eq. (6.19), we obtain:

$$L_f \, dL_f = K\frac{\Delta H_p}{\Delta\theta} \, dt = \tilde{D} \, dt \tag{6.20}$$

where the composite term $(K \, \Delta H_p / \Delta\theta)$ can be regarded as an effective diffusivity $\tilde{D}$ for the infiltrating profile. Integration gives

$$\frac{L_f^2}{2} = K\frac{\Delta H_p}{\Delta\theta} t = \tilde{D}t \tag{6.21}$$

$$L_f = \sqrt{2Kt \, \Delta H_p/\Delta\theta} = \sqrt{2\tilde{D}t} \tag{6.22}$$

or

$$I = \Delta\theta\sqrt{2\tilde{D}t}, \qquad i = \Delta\theta\sqrt{\tilde{D}/2t} \tag{6.23}$$

which compares with Eqs. (6.4) and (6.5) (the difference being in the $\sqrt{2\pi}$ ratio for the weighting of $\bar{D}$ vs. $\tilde{D}$, both being approximate[7]). Thus the depth of the wetting front is proportional to $\sqrt{t}$, and the infiltration rate is proportional to $1/\sqrt{t}$.

With gravity taken into account, the Green and Ampt approach gives

$$\frac{dI}{dt} = \Delta\theta \frac{dL_f}{dt} = K\frac{H_0 - H_f + L_f}{L_f} \tag{6.24}$$

which integrates to

$$\frac{Kt}{\Delta\theta} = L_f - (H_0 - H_f) \ln\left(1 + \frac{L_f}{H_0 - H_f}\right) \tag{6.25}$$

As t increases, the second term on the right-hand side of Eq. (6.25) increases more and more slowly in relation to the increase in L_f, so that, at very large times, we can approximate the relationship by

---

[7] $\tilde{D}$ can be regarded as an indication of what wetting-front value must be assumed for the Green and Ampt approach to work.

$$L_f \cong \frac{Kt}{\Delta\theta} + \delta \tag{6.26}$$

or

$$I \cong Kt + \delta$$

where δ can eventually be regarded as a constant.

The Green and Ampt relationships are essentially empirical, since the value of the effective wetting-front suction must be found by experiment. For infiltration into initially dry soil, it may be of the order of -50 to -100 cm H_2O, or ~ -0.1 bar (Green and Ampt, 1911; Hillel and Gardner, 1970). However, in actual field conditions, particularly where the initial moisture is not uniform, H_f may be undefinable. In many real situations, the wetting front is too diffuse to indicate its exact location at any particular time.

G. Infiltration into Layered Soils

The effect of profile stratification on infiltration was studied by Hanks and Bowers (1962),[8] who used a numerical technique for analyzing the flow equation, and by Miller and Gardner (1962), who conducted experiments on the effect of thin layers sandwiched into otherwise uniform profiles. A conducting soil must have continuous matric suction and hydraulic-head values throughout its length, regardless of layering sequence. However, the wetness and conductivity values may exhibit abrupt discontinuities at the interlayer boundaries.

One typical situation is that of a coarse layer of higher saturated hydraulic conductivity, overlying a finer-textured layer. In such a case, the infiltration rate is at first controlled by the coarse layer, but when the wetting front reaches and penetrates into the finer-textured layer, the infiltration rate can be expected to drop and tend to that of the finer soil alone. Thus, in the long run, it is the layer of lesser conductivity which controls the process. If infiltration continues for long, then positive pressure heads (a "perched water table") can develop in the coarse soil, just above its boundary with the impeding finer layer.

In the opposite case of infiltration into a profile with a fine-textured layer over a coarse-textured one, the initial infiltration rate is again determined by the upper layer. As water reaches the interface with the coarse lower layer, however, the infiltration rate may decrease. Water at the wetting front is normally under suction, and this suction may be too high to permit entry into the relatively large pores of the coarse layer. This explains the observation

---

[8] This technique was used by Green et al. (1962) to estimate infiltration in the field.

(Miller and Gardner, 1962) that the wetting-front advance stops for a time (though infiltration at the surface does not stop) until the pressure head at the interface builds up sufficiently to penetrate into the coarse material. Thus, a layer of sand or gravel in a medium or fine-textured soil, far from enhancing water movement in the profile, may actually impede it. The lower layer, in any case, cannot become saturated, since the restricted rate of flow through the less permeable upper layer cannot sustain flow at the saturated hydraulic conductivity of the coarse lower layer (except when the externally applied pressure, i.e., the ponding depth, is large).

The steady-state downflow of water through a two-layer profile into a free-water table beneath was analyzed by Takagi (1960). Where the upper layer is less pervious than the lower, negative pressures (suctions) were shown to develop in the lower layer, and these can remain constant throughout a considerable depth range.

H. Infiltration into Crust-Topped Soils

A very important special case of a layered soil is that of an otherwise uniform profile which develops a crust, or seal, at the surface. Such a seal can develop under the beating action of raindrops (Ekern, 1950; McIntyre, 1958; Tackett and Pearson, 1965), or as a result of the spontaneous slaking and breakdown of soil aggregates during wetting (Hillel, 1960). Surface crusts are characterized by greater density, finer pores, and lower saturated conductivity than the underlying soil. Once formed, a surface crust can greatly impede water intake by the soil (Fig. 6.6), even if the crust is quite

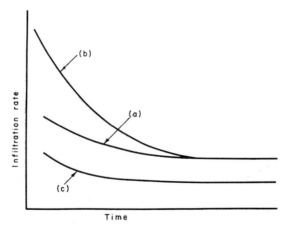

Fig. 6.6. Infiltrability as a function of time: (a) in a uniform soil; (b) in a soil with a more porous upper layer; and (c) in a soil covered by a surface crust.

thin (say, not more than several millimeters in thickness) and the soil is otherwise highly permeable. Failure to account for the formation of a crust can result in gross overestimation of infiltration.

An analysis of the effect of a developing surface crust upon infiltration was carried out by Edwards and Larson (1969), who adapted the Hanks and Bowers (1962) numerical solution to this problem. Hillel (1964), and Hillel and Gardner (1969, 1970) used a quasianalytical approach to calculate fluxes during steady and transient infiltration into crust-capped profiles from knowledge of the basic hydraulic properties of the crust and of the underlying soil.

The problem is relatively simple in the case of steady infiltration. Steady-state conditions require that the flux through the crust q_c be equal to the flux through the subcrust "transmission zone" q_u:

$$q_c = q_u$$

or

$$K_c \left(\frac{dH}{dz}\right)_c = K_u \left(\frac{dH}{dz}\right)_u \tag{6.27}$$

where K_c, $(dH/dz)_c$, K_u, and $(dH/dz)_u$ refer to the hydraulic conductivity and hydraulic-head gradient of the crust and underlying transmission zone, respectively. The gradient through the transmission zone tends to unity when steady infiltration is approached, as the suction gradient decreases with the increase in wetting depth, eventually leaving the gravitational gradient as the only effective driving force. In the absence of a suction-head gradient in the zone below the crust, we obtain (with the soil surface as our reference level)

$$q = K_u(\psi_u) = K_c \frac{H_0 + \psi_u + z_c}{z_c} \tag{6.28}$$

where $K_u(\psi_u)$ is the unsaturated hydraulic conductivity of the subcrust zone, a function of the suction head ψ_u which develops in this zone, beginning just under the hydraulically impeding crust; H_0 is the positive hydraulic head imposed on the surface by the ponded water; and z_c is the vertical thickness of the crust.

Where the ponding depth H_0 is negligible and the crust itself is very thin and of low conductivity (e.g., where z_c is very small in relation to the suction ψ_u which forms at the subcrust interface), we can assume the approximation

$$q_u = q_c = K_c \frac{\psi_u}{z_c} \tag{6.29}$$

The condition that the crust remain saturated even while its lower part will be under suction is that its critical air-entry ψ_a not be exceeded (i.e., $\psi_u < \psi_a$).

This, together with the condition that the subcrust hydraulic-head gradient approximate unity, leads to the approximation

$$\frac{K_u}{\psi_u} = \frac{K_c}{z_c} = \frac{1}{R_c} \tag{6.30}$$

i.e., the ratio of the hydraulic conductivity of the underlying soil transmission zone to its suction is approximately equal to the ratio of the crust's (saturated) hydraulic conductivity to its thickness. The latter ratio is the reciprocal of the hydraulic resistance per unit area of the crust R_c.[9] Also, by Eq. (6.28),

$$q = K_u(\psi_u) = \psi_u/R_c \tag{6.31}$$

Where the unsaturated conductivity of the underlying soil bears a known single-valued relation to the suction, it should be possible to calculate the steady infiltration rate and the suction in the subcrust zone on the basis of the measurable hydraulic resistance of the crust. Where the relation of matric suction to water content is also known, it should be possible to infer the subcrust water content during steady infiltration.

Employing a K vs. ψ relationship of the type $K = a\psi^{-n}$ (where a, and n are characteristic constants of the soil), Hillel and Gardner (1969) obtained the following[10]:

$$q = \frac{a^{1/(n+1)}}{R_c^{n/(n+1)}} = \frac{B}{R_c^{n/(n+1)}} \tag{6.32}$$

$$\psi_u = (aR_c)^{1/(n+1)} = BR_c^{1/(n+1)} \tag{6.33}$$

where $B = a^{1/(n+1)}$ is a property of the subcrust soil. The theoretical consequences of Eqs. (6.32) and (6.33) are illustrated in Fig. 6.7. These equations indicate how the infiltration rate decreases, and the subcrust suction increases, with increasing hydraulic resistance of the crust. Gardner (1956) has shown that the values of a and of n generally increase with increasing coarseness, textural as well as structural, of the soil. Sands may have n values of four or more, whereas clayey soils may have n values of about two. Tillage may pulverize and loosen the soil, thus increasing n, whereas compaction may have the opposite effect.

Both the crust and the underlying soil are seen to affect the infiltration rate and suction profile, and the crust-capped soil is thus viewed as a self-adjusting system in which the physical properties of the crust and underlying

---

[9] A distinction is made between the hydraulic resistance per unit area, defined as above, and the hydraulic resistivity, the latter being equal to the reciprocal of the conductivity.

[10] The relation of conductivity to suction does not always obey so simple an equation as $K = a\psi^{-n}$. An alternative expression, proposed by Hillel and Gardner (1969), may have more general validity: $K = K_s(\psi_a/\psi)^n$, for $\psi > \psi_a$.

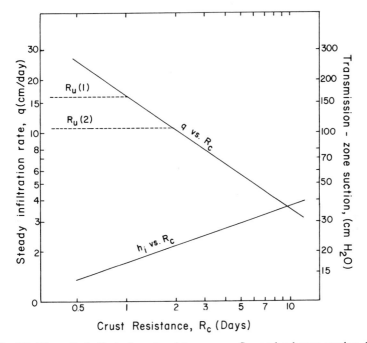

Fig. 6.7. Theoretical effect of crust resistance upon flux and subcrust suction during steady infiltration into crust-capped columns of a uniform soil with $n = 2$, $a = 4.9 \times 10^3$. The broken lines (1) and (2) indicate the hypothetical effect of subcrust hydraulic resistance R_u: $R_u(1) < R_u(2)$. The decreasing q vs. R_c curve applies only where the hydraulic conductance of the subcrust layers is not limiting. (After Hillel and Gardner, 1969.)

soil interact in time to form a steady infiltration rate and moisture profile. In this steadily infiltrating profile, the subcrust suction which develops is such as to create a gradient through the crust and a conductivity in the subcrust zone which will result in an equal flux through both layers.

The problem is rather more complicated in the prevalent case of transient infiltration into an initially unsaturated profile, during which the flux, the wetting depth, the subcrust suction, and the conductivity might all be changing with time.

Assuming the Green and Ampt conditions (Section 6F), and with H_0 negligible, Hillel and Gardner (1970) recognized three stages during transient infiltration into crusted profiles: an initial stage, in which the rate is finite and dependent on crust resistance R_c and on an effective subsoil suction; an intermediate stage, in which cumulative infiltration I increases approximately as the square root of time; and a later stage, in which I can be expressed as the sum of a steady and a transient term, the latter becoming negligible at long times. I was shown to decrease with increasing R_c, particularly in coarse-textured and coarse-structured soils. Experimental data indicate that the cumulative infiltration curves of crusted profiles scale as the square root of their transmission-zone diffusivities. Thus, infiltration into a crusted profile can be described by the approximation that water enters into the subcrust soil at a nearly constant suction, the magnitude of which is determined by crust resistance and hydraulic characteristics of the soil.

Where the gravity effect is negligible (e.g., in horizontal flow or during the initial stages of vertical infiltration into an initially dry medium of high matric suction), the infiltration vs. time relationship was given by:

$$I = \sqrt{K_u^2 R_c^2 (\Delta\theta)^2 + 2K_u H_f\, \Delta\theta\, t} - K_u R_c \qquad (6.34)$$

Where the gravity effect is significant, the expression given is

$$L_f = \frac{K_u t}{\Delta\theta} + (H_f - K_u R_c)\ln\left[\frac{H_f + (K_u t/\Delta\theta) + \delta(t)}{H_f}\right] \qquad (6.35)$$

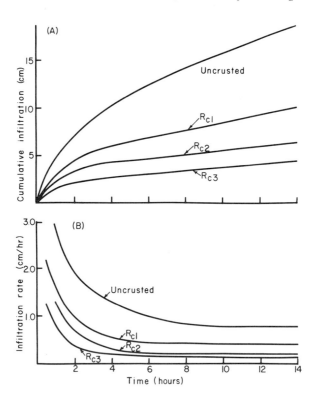

Fig. 6.8. Time dependence of cumulative infiltration (A) and of infiltration rate (B) for uncrusted and crusted columns of Negev loess. Crust resistance values R_{c1}, R_{c2}, R_{c3} are 3.2, 9.1, and 17 days, respectively (after Hillel and Gardner, 1969).

where the correction term $\delta(t)$ becomes negligibly small as t increases. Thus, L_f can be expressed as the sum of a steady and a transient term. Some experimental results are shown in Fig. 6.8.

I. Rain Infiltration

When rain or sprinkling intensity exceeds soil infiltrability, the infiltration process is the same as in the case of shallow ponding. If rain intensity is less than the initial infiltrability value of the soil, but greater than the final value, then at first the soil will absorb water at less than its potential rate and flow in the soil will occur under unsaturated conditions; however, if the rain is continued at the same intensity, and as soil infiltrability decreases, the soil surface will eventually become saturated and henceforth the process will continue as in the case of ponding infiltration. On the other hand, if rain intensity is at all times lower than soil infiltrability (i.e., lower than the saturated hydraulic conductivity), the soil will continue to absorb the water as fast as it is applied without ever reaching saturation. After a long time, as the suction gradients become negligible, the wetted profile will attain a wetness for which the conductivity is equal to the water application rate, and the lower this rate, the lower the degree of saturation of the infiltrating profile. This effect is illustrated in Fig. 6.9.

The process of infiltration under rain or sprinkler irrigation was studied by Youngs (1960) and by Rubin and Steinhardt (1963, 1963), Rubin *et al.* (1964), and Rubin (1966). The latter author, who used a numerical solution of the flow equation for conditions pertinent to this problem, recognized three modes of infiltration due to rainfall: (1) *nonponding infiltration*, involving rain not intense enough to produce ponding; (2) *preponding infiltration*, due to rain that can produce ponding but that has not yet done so; and (3) *rainpond infiltration*, characterized by the presence of ponded water. Rainpond infiltration is usually preceded by preponding infiltration, the transition between the two being called *incipient ponding*. Thus, nonponding and preponding infiltration are *rain-intensity-controlled* (or *flux-controlled*), whereas rainpond infiltration is controlled by the pressure (or depth) of water above the soil surface, as well as by the suction conditions and conductivity relations of the soil. Where the pressure at the surface is small, rainpond infiltration, like ponding infiltration in general, is *profile-controlled*.

In the analysis of rainpond or ponding infiltration, the surface boundary condition generally assumed is that of a constant pressure at the surface, whereas in the analysis of nonponding and preponding infiltration, the water flux through the surface is considered to be constant, or increasing. In actual field conditions, rain intensity might increase and decrease alternately, at times exceeding the soil's saturated conductivity (and its infiltrability) and

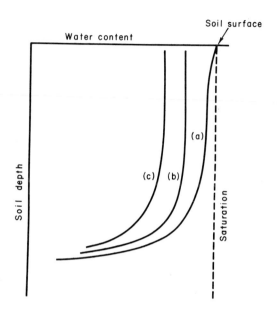

Fig. 6.9. The water-content distribution profile during infiltration: (a) under ponding; (b) under sprinkling at relatively high intensity; and (c) under sprinkling at a very low intensity.

at other times dropping below it. However, since periods of decreasing rain intensity involve complicated hysteresis phenomena, the analysis of composite rainstorms is very difficult and has not yet been carried out satisfactorily.

Rubin's analysis is based on the assumption of no hysteresis. The falling raindrops are taken to be so small and numerous that rain may be treated as a continuous body of "thin" water reaching the soil surface at a given rate. Soil air is regarded as a continuous phase, at atmospheric pressure. The soil is assumed to be uniform and stable (i.e., no fabric changes such as surface crusting).

We shall briefly review the consequences of Rubin's analysis in qualitative terms. If a constant pressure head is maintained at the soil surface (as in rainpond infiltration), then the flux of water into this surface must be constantly decreasing with time. If a constant flux is maintained at the soil surface, then the pressure head at this surface must be constantly increasing with time. Infiltration of constant-intensity rain can result in ponding only if the *relative rain intensity* (i.e., the ratio of rain intensity to the saturated hydraulic conductivity of the soil) exceeds unity. During nonponding infiltration under a constant rain intensity q_r, the surface suction will tend to a limiting value ψ_{lim} such that $K(\psi_{\mathrm{lim}}) = q_r$.

Under rainpond infiltration, the wetted profile consists of two parts: an upper, water-saturated part; and a lower, unsaturated part. The depth of the saturated zone continuously increases with time. Simultaneously, the steepness of the moisture gradient at the lower boundary of the saturated zone (i.e., at the wetting zone and the wetting front) is continuously decreasing (these phenomena accord with those of infiltration processes under ponding, as described in the previous sections of this chapter). The higher the rain intensity is, the shallower is the saturated layer at incipient ponding and the steeper is the moisture gradient in the wetting zone.

Figure 6.10 describes infiltration rates into a sandy soil during preponding

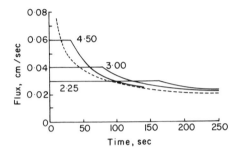

Fig. 6.10. Relation between surface flux and time during infiltration into Rehovot sand due to rainfall (solid lines) and flooding (dashed line). The numbers labeling the curves indicate the magnitude of the relative rain intensity (after Rubin, 1966).

and rainpond infiltration under three rain intensities. The horizontal parts of the curves correspond to preponding infiltration, and the descending parts to rainpond infiltration periods. As pointed out by Rubin (1966), the rain-pond infiltration curves are of the same general shape and approach the same limiting infiltration rate, but they do not constitute horizontally displaced parts of a single curve, and do not coincide with the infiltration rate under flooding, which is shown as a broken line in the same graph.

The process of rain infiltration has not yet been studied in sufficient detail in the field to establish the applicability of existing theories. Complications due to the discreteness of raindrops (which causes alternate saturation and redistribution at the surface), as well as to the highly variable nature of rain-storm intensities and raindrop energies, and the unstable nature of many (perhaps even most) soils, can cause anomalies disregarded by idealized theories. Additional complications can arise in cases of air occlusion and when the soil exhibits profile or areal heterogeneity.

J. Surface Runoff

Surface runoff, or overland flow, is the portion of the rain which is not absorbed by the soil and does not accumulate on the surface, but runs down-slope and collects in gullies and streams. Runoff can occur only when rain intensity exceeds the infiltration rate. Even then, however, runoff does not begin immediately, as the excess rain first collects in surface depressions and forms puddles, whose total volume is termed the *surface storage capacity*. Only when the surface storage is filled and the puddles begin to overflow does runoff begin. The rate of the runoff flow depends upon the excess of rain intensity over the infiltration rate. Obviously, the surface storage also depends on the slope, as well as on the roughness of the soil surface.

In agricultural fields, runoff is generally undesirable, since it results in loss of water and often causes erosion, the amount of which increases with increasing rate and velocity of runoff. The way to prevent erosion is to protect the soil surface against raindrop splash (e.g., by mulching), to increase soil infiltrability and surface storage, and to obstruct overland flow so as to prevent it from gathering velocity. Maintenance and stabilization of soil aggregation will minimize slaking and detachment of soil particles by raindrops and running water. A crusted or compacted soil generally has a low infiltration rate and therefore will produce a high rate of runoff. Proper tillage, especially on the contour, can increase infiltration and surface storage capacity, thus reducing runoff (Burwell and Larson, 1969).

In arid regions, it is sometimes desirable to induce runoff artificially in order to supply water for human, industrial, or agricultural use (Hillel *et al.*, 1967).

K. Summary

An important physical property of a soil is the rate at which it can absorb water supplied to its surface. This rate, termed *soil infiltrability*, depends on the following factors:

(1) Time from the onset of the rain or irrigation: The infiltration rate is apt to be relatively high at first, then to decrease, and eventually to approach a constant rate that is characteristic for the soil.

(2) Initial water content: The wetter the soil is initially, the lower will be the initial infiltrability (owing to smaller suction gradients) and the quicker will be the attainment of the final (constant) rate, which itself is generally independent of the initial water content.

(3) Hydraulic conductivity: The higher the saturated hydraulic conductivity of the soil is, the higher its infiltrability tends to be.

(4) Soil surface conditions: When the soil surface is highly porous and of "open" structure, the initial infiltrability is greater than that of a uniform

soil, but the final infiltrability remains unchanged, as it is limited by the lower conductivity of the transmission zone beneath. On the other hand, when the soil surface is compacted and the profile covered by a surface crust of lower conductivity, the infiltration rate is lower than that of the uncrusted (uniform) soil. The surface crust acts as a hydraulic barrier, or bottleneck, impeding infiltration. This effect, which becomes more pronounced the thicker and the denser the crust, reduces both the initial and the final infiltration rate. A soil of unstable structure tends to form such a crust during infiltration, especially as the result of the slaking action of beating raindrops. In such a soil, a plant cover or a surface mulch of plant residues can serve to intercept and break the impact of the raindrops and thus help to prevent crusting.

(5) The presence of impeding layers inside the profile: Layers which differ in texture or structure from the overlying soil may retard water movement during infiltration. Perhaps surprisingly, clay layers and sand layers can have a similar effect, although for opposite reasons. The clay layer impedes flow owing to its lower *saturated* conductivity, while a sand layer retards the wetting front (where unsaturated conditions prevail) owing to the lower *unsaturated* conductivity of the sand. Flow into a dry sand layer can take place only after the pressure head has built up sufficiently for water to move into and fill the large pores of the sand.

7 Redistribution of Soil Moisture Following Infiltration

A. General

When rain or irrigation ceases, and surface storage is depleted by evaporation or infiltration, the infiltration process comes to an end, since no more water enters into the soil. Downward water movement within the soil, however, does not cease immediately, and may in fact persist for a long time as soil water redistributes within the profile. The soil layer wetted to near-saturation during infiltration does not retain its full water content, since part of its water moves down into the lower layers under the influence of gravity and possibly also of suction gradients. In the presence of a high water table, this post-infiltration movement is sometimes termed *internal drainage*. In the absence of groundwater, or where the water table is too deep to affect the relevant depth zone, this movement is called *redistribution*, since its effect is to redistribute soil water by increasing the wetness of successively deeper layers within the profile at the expense of the initially wetted upper layer.

In some cases, the rate of redistribution decreases rapidly, becoming imperceptibly small after several days so that thereafter the wetted part of the soil appears to retain its moisture, unless this moisture is evaporated or is taken up by plants. In other cases, redistribution may continue at an appreciable, though diminishing, rate for many days and even weeks.

The importance of the redistribution process should be self-evident, as it determines the amount of water retained at various times by the different depth zones in the soil profile, and hence it can affect the water economy of

155

plants. The rate and duration of downward flow during redistribution determine the effective *water-storage capacity* of the soil, a property that is vitally important, particularly in relatively dry regions, where water supply is infrequent and plants must rely for long periods on the unreplenished reservoir of water within the rooting zone. As we shall explain more fully in the sections to follow, soil-water storage is generally not a fixed quantity or a static property but a temporary phenomenon, determined by the dynamics of soil-water flow.

B. Description of the Process

We have already made a distinction between the post-infiltration movement of soil water in cases where a groundwater table is present fairly close to the soil surface (i.e., at a depth not exceeding a few meters), and in cases where groundwater is either nonexistent or too deep to affect the state and movement of soil moisture in the root zone.

At the groundwater level, also called the *phreatic surface*, soil water is at atmospheric pressure. Beneath this level, the hydrostatic pressure exceeds atmospheric pressure, while above this level, soil water is under suction. Internal drainage in the presence of a water table tends to a state of equilibrium, in which the suction at each point corresponds to its height above the free water level. We shall discuss internal drainage and groundwater flow phenomena in our next chapter.

In the absence of groundwater, and provided the soil is sufficiently deep, the typical profile at the end of the infiltration process consists of a wetted zone in the upper part of the profile, and an unwetted zone beneath. Such a profile is like a bottomless barrel, with the deeper layers (those beyond the infiltration wetting front) drawing water from the upper ones. The initial rate of redistribution depends on the initial wetting depth, as well as on the relative dryness of the bottom layers and the hydraulic properties of the conducting soil. When the initial wetting depth is small, and when the underlying soil is relatively dry, the suction gradients are likely to be strong and the redistribution rate rapid. On the other hand, when the initial wetting depth is considerable, and when the underlying soil is wet, the suction gradients are small and redistribution occurs primarily under the influence of gravity.

In any case, the rate of redistribution generally decreases in time, for two reasons: (1) the suction gradients between the wet and dry zones decrease as the former loses, and the latter gains, moisture; (2) as the initially wetted zone desorbs, its hydraulic conductivity decreases correspondingly. With both gradient and conductivity decreasing simultaneously, the flux falls rapidly. The rate of advance of the wetting front decreases accordingly, and

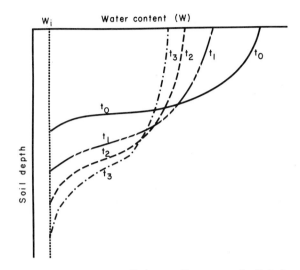

Fig. 7.1. The changing moisture profile in a medium-textured soil during redistribution following an irrigation. The moisture profiles shown are for 0, 1, 4, and 14 days after the irrigation. w_i is preirrigation soil wetness.

this front, which was relatively sharp during infiltration, gradually dissipates during redistribution. This is illustrated in Fig. 7.1.

The figure shows that the upper, initially-wetted zone drains monotonically, though at a decreasing rate. On the other hand, the sublayer at first

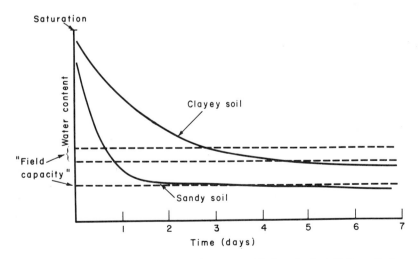

Fig. 7.2. The monotonic decrease of soil wetness with time, in the initially wetted zone, during redistribution.

wets up, but eventually begins also to drain. The time dependence of soil wetness in the upper zone is illustrated in Fig. 7.2 for a sandy soil, in which the unsaturated conductivity falls off rapidly with increasing suction, and for a clayey soil, in which the decrease of conductivity is more gradual and redistribution persists longer.

C. Hysteretic Phenomena in Redistribution

The redistribution process involves hysteresis. Since the upper part of the profile is desorbing (draining) while the lower part is sorbing water, the relation between wetness and suction will be different in different depths, and may change with time, even in a profile of uniform texture. We shall recall from Chapter 3 that the relation between wetness and suction is not unique but depends on the history of wetting and drying that take place at each point in the soil. This relationship, when plotted, exhibits two limiting curves which apply when wetting or drying starts from the extreme conditions of dryness or saturation, respectively. Between the wetting and drying branches there lies an infinite number of possible "scanning curves" which describe wetting or drying between different intermediate values of wetness. In general, the effect of hysteresis is to retard redistribution.

The hysteresis phenomenon complicates the redistribution process and makes it difficult to describe and analyze mathematically. To do this, it is necessary to know the wetting history and the wetness value when desorption commences at each depth within the profile. Furthermore, it is necessary to know the slopes of all possible scanning curves (these slopes, in general, differ considerably from those of both the primary-branch curves). Hence, it appears at present that the best way to handle hysteresis is by numerical analysis (e.g., Rubin, 1967), while analytical or quasianalytical attempts to treat the redistribution process have so far tended to avoid hysteresis or to assume rather gross simplifications (e.g., Youngs, 1958a,b; Gardner et al., 1970).

D. Analysis of Redistribution Processes

The equation for flow in a vertical profile is

$$\frac{\partial \theta}{\partial t} = -\frac{\partial}{\partial z}\left(K\frac{\partial \psi}{\partial z} + K\right) \tag{7.1}$$

In the case of redistribution involving hysteresis, this equation may be written (Miller and Klute, 1967)

$$\left(\frac{d\theta}{d\psi}\right)_h \frac{\partial \psi}{\partial t} = -\frac{\partial}{\partial z}\left[K_h(\psi)\frac{\partial \psi}{\partial z}\right] - \frac{\partial K_h(\psi)}{\partial z} \tag{7.2}$$

In the above, θ is volumetric wetness, t time, K conductivity, z depth, ψ suction head, and the subscript h indicates a hysteretic function. After the cessation of infiltration, and in the absence of evaporation, flux through the soil surface is zero and hence the hydraulic gradient at the surface must also be zero. Conservation of matter requires that

$$\int_{z=0}^{z=\infty} \theta\, dz = \text{const} \tag{7.3}$$

for all time, provided no sinks are present (e.g., no extraction of water by plant roots).

When redistribution begins, the upper portion of the profile, which was wetted to near-saturation during the preceding infiltration process, begins to desorb monotonically. Below a certain depth, however, the soil first wets during redistribution, then begins to drain, and the value of wetness at which this turnabout takes place decreases with depth. Each point in the soil thus follows a different scanning curve, and the conductivity and water-capacity functions vary with position. An approximate analysis of these phenomena was given by Youngs (1958a,b; 1960). He showed that in coarse materials [with uniform pore sizes and hence with "abrupt" $\theta(\psi)$ curves] redistribution can be slower for shallow than for deep wettings.

The flow equation describing redistribution was analyzed numerically by Staple (1966, 1969), Remson et al. (1965, 1967), and Rubin (1967). Rubin's analysis was based on the assumptions that, despite hysteresis, there exists at any time and depth in the soil a unique, single-valued relation between wetness and suction; that a soil element that has begun to desorb will not wet up again; and that hysteresis in the relation of conductivity to wetness (but not to suction) can be neglected.

To solve Eq. (7.1), one of the two dependent variables (namely, ψ) was eliminated:

$$\frac{\partial \theta}{\partial t} = -\frac{\partial}{\partial z}\left(\frac{K}{c}\frac{\partial \theta}{\partial z} + K\right) \tag{7.4}$$

The solution was then sought, subject to the following initial and boundary conditions:

$$t = 0, \qquad z > 0, \qquad \theta = \theta(z)$$

$$t > 0, \qquad z = 0, \qquad q = \frac{K}{c}\frac{\partial \theta}{\partial z} + K = 0$$

$$t > 0, \qquad z = \infty, \qquad \theta = \theta_i \tag{7.5}$$

where q is the flux, $\theta(z)$ is a function of soil depth describing the post-infiltration (pre-redistribution) moisture profile, θ_i is the soil's pre-infiltration wetness, c is the differential water capacity $d\theta/d\psi$, and the other variables are as defined above. The ratio K/c is the diffusivity D.

Equation (7.4) was approximated by an implicit difference equation. In order to take hysteresis into account, empirical equations were used to describe the dependence of ψ on θ in primary wetting, in primary drying, and in intermediate (transitional) scanning from wetting to drying. Another empirical equation was used to describe the relation of K to θ. Empirical equations cannot be expected to pertain to any but the particular soil considered. The results of the analysis, however, are thought to be generally valid in principle, provided the basic assumptions are met in the real situation, at least approximately.

Rubin's findings indicate that the hysteretic moisture profile is not bounded by the two possible nonhysteretic profiles (one assuming the desorbing branch of the soil-moisture characteristic, the other assuming the sorbing branch). The hysteretic redistribution was shown to be clearly slower than the nonhysteretic one. These results demonstrate the importance of hysteresis in redistribution processes, particularly in coarse-textured soils.

A different approach to the analysis of redistribution was taken by Gardner $et\ al.$ (1970). Equation (7.4) can be solved analytically by separation of variables (Gardner, 1962) in the special case where the empirical relation $D = C\theta^n$ applies (with C and n constants). In this procedure, it is assumed that the solution is of the form $\theta = T(t)Z(z)$, where T is a function of t alone and Z is a function of z alone. It is further assumed that $K = B\theta^m$ (with B and m constants). Their analysis suggests that both where matric suction forces predominate and where, alternatively, the gravity force predominates the time dependence of soil wetness (and of total water content in any depth zone) obeys a relation of the type

$$\bar{\theta} \quad \text{or} \quad \int \theta dz \approx a(b+t)^{-c} \tag{7.6}$$

where $\bar{\theta}$ is the mean wetness of the initially-wetted zone at time t during redistribution. This is shown in Fig. 7.3.

An alternative approximate analysis which provides insight into the redistribution process and permits useful predictions was carried out by Peck (1970). He introduced two assumptions:

(1) the envelope of all instantaneous moisture profiles can be represented by a rectangular hyperbola for soil wetness values lower than the maximum wetness attained during infiltration; and

(2) soil wetness at the interface between the desorbing and the sorbing parts of the profile is approximately a linear function of the desorbing zone above this interface. With these assumptions, a first order ordinary differential equation is derived which can be integrated numerically to relate the mean wetness of the desorbing zone to the redistribution time.

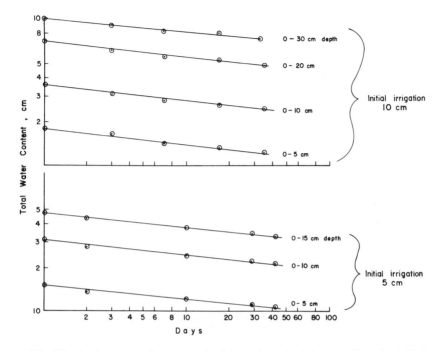

Fig. 7.3. Total amount of water retained in various depth layers within the initially wetted zone of a fine, sandy loam during redistribution following irrigations of 5 and 10 cm. of water (after Gardner *et al.*, 1970).

Peck defined the mean wetness in the desorbing zone $0 \leq z < z_*$ by

$$\bar{\theta} = (1/z_*) \int_0^{z*} \theta \, dz \tag{7.7}$$

where z_* is the depth of the transition plane at any particular time. He showed that with increasing infiltration quantity, I, both the depth of the transition plane and its rate of downward movement are increased. However, the flux of water through the transition plane is reduced. Neglecting the influence of gravity, the time needed for mean soil wetness to decrease to any specified value were shown to increase with I^2. Analysis of the possible effect of initial (pre-infiltration) soil wetness indicated that the higher initial wetness leads immediately to greater penetration of the infiltrating water, but since the overall gradient is reduced, water movement through the transition plane occurs less rapidly. Although mean soil wetness is always higher in the initially-wetter soil, the moisture increment $(\theta - \theta_{initial})$ is greater in the initially-drier case.

E. "Field Capacity"

Early observations that the rate of flow and water-content changes decrease in time (Alway and McDole, 1917; Richards and Moore, 1952; Veihmeyer and Hendrickson, 1931) have been construed to indicate that the flow rate generally becomes negligible within a few days, or even that the flow ceases entirely. The presumed water content at which internal drainage allegedly ceases, termed the *field capacity*, had for a long time been accepted almost universally as an actual physical property, characteristic of and constant for each soil.

Though the field-capacity concept originally derived from rather crude measurements of water content in the field (where sampling and measurement errors necessarily limited the accuracy and validity of the results), some workers have sought to explain it somehow in terms of a static equilibrium value or a discontinuity in the capillary water. It was commonly assumed that the application of a certain quantity of water to the soil will fill the deficit to field capacity to a certain and definite depth, beyond which this quantity of water will not penetrate. It became an almost universal practice to calculate the amount of irrigation to be applied at any particular time on the basis of the deficit to field capacity of the soil depth layer to be wetted.

In recent years, with the development of theory and more precise experimental techniques in the study of unsaturated flow processes, the field-capacity concept, as originally defined,[1] has been recognized as arbitrary and not an intrinsic physical property independent of the way it is measured. The field-capacity concept may actually have "done more harm than good" (Richards, 1960). When and how can one determine that redistribution has "materially decreased," "virtually ceased," or that its rate has become "negligible" or "practically zero"? Obviously, the criteria for such a determination are subjective, depending as often as not upon the frequency and accuracy with which the soil water content is measured. The common working definition of field capacity (namely, the wetness of the initially wetted zone, say, two days after infiltration) takes no account of such factors as the antecedent (pre-infiltration) wetness of the soil, the depth of wetting, or the amount of irrigation applied, etc.

The redistribution process is in fact continuous, and exhibits no abrupt "breaks" or static levels. Although its rate decreases constantly, in the absence of a water table the process continues and equilibrium is approached, if at all, only after very long periods.

[1] According to Veihmeyer and Hendrickson (1949), the field capacity is "the amount of water held in soil after excess water has drained away and the rate of downward movement has materially decreased, which usually takes place within 2–3 days after a rain or irrigation in pervious soils of uniform structure and texture."

The soils for which the field-capacity concept is most tenable are the coarse-textured ones, in which the hydraulic conductivity drops most steeply with increasing matric suction, and flow becomes slow relatively soon. In medium-textured or fine-textured soils, however, redistribution can persist at an appreciable rate for many days. As an example, we can cite the case of a loessial silt loam in the Negev region of Israel, in which the changes in water content shown in Table 7.1 were observed in the 60–90-cm depth

Table 7.1

	Per cent moisture content by mass
At the end of infiltration	29
After 1 day	20.2
2 days	18.7
7 days	17.5
30 days	15.9
60 days	14.7
156 days	13.6

zone following a wetting to a depth exceeding 150 cm (with evaporation prevented by means of a paper mulch). It is seen that the water loss continued incessantly for over five months. The rate of decrease of water content obeyed the function (Richards *et al.*, 1956):

$$\frac{dw}{dt} = at^{-b} \tag{7.8}$$

where w is the water content, t the time, and a and b are constants related to the boundary conditions and conductance properties of the soil. The exponential constant b, which is related to soil diffusivity, is obviously most important, and the greater its value, the steeper the decrease in water content. It is therefore more plausible to assume that water retention in the soil is related to soil diffusivity, or to hydraulic conductivity, rather than to suction *per se*.

The rate of outflow from any given layer in the soil depends not only on the texture or hydraulic characteristics of that layer, but also on the composition and structure of the entire profile, since the presence at any depth of an impeding layer can retard the movement of water out of the layers above it. Thus, it is clear that the storage capacity of the soil is related not only to time, but also to the textural composition and layering sequence of the profile, as well as to its initial water-content distribution.

An agriculturist engaged in irrigated farming and accustomed to frequent irrigations is interested mainly in the short-run storage capacity of his soil. For him, the field capacity of the loessial soil cited can be taken at about 18%. By way of contrast, an agriculturist engaged in dryland farming is sometimes interested in storing soil water from one season, or even from one year, to the next. For the dryland farmer, therefore, the field capacity of the same loessial soil cannot be taken at 18% (since the soil does not retain this content beyond a few days), but at 14% or even less.

Our assumption thus far has been that the redistribution process occurs by itself—that is, in the absence of other processes of water extraction from the soil. This assumption is seldom, if ever, realistic. When several processes of soil-water extraction (e.g., internal drainage, evaporation, and uptake by plants) occur simultaneously, the reduction in soil water content is obviously more rapid, and less apt to cease at any point such as "field capacity."

Despite all these objections, the field-capacity concept is still considered by many to be a useful practical criterion for the upper limit of soil water content which can be depended upon (more or less) in the field. As such, field capacity must be measured in the field, and there is no generally satisfactory laboratory procedure for obtaining it. The value of the various laboratory methods which have nevertheless been proposed (e.g., equilibration of soil samples in a centrifuge or in a suction apparatus) depends on the degree to which their results correlate with the field measurement taken under standard conditions. Such suction values as 1/3 or 1/10 bar may represent a measurable field capacity in certain circumstances, but it is fundamentally wrong to expect or assume that such criteria will hold universally, since they are solely static in nature while the redistribution process is essentially dynamic. The important fact is that, although field capacity is indefinite, the rapid slowing of redistribution is an exceedingly valuable property of soils generally, as it results in the storage (albeit temporary) of water for plant use during periods of no rain or irrigation.

F. Factors Affecting Redistribution and "Field Capacity"

The field capacity, as commonly measured, may vary between about 4% (by mass) in sands, to about 45% in heavy clay soils, and up to 100% or even more in certain organic soils. Among the factors affecting redistribution and the apparent field capacity are the following:

(1) Soil texture: Clayey soils retain more water, and longer, than sandy ones. Hence, the finer the texture is, the higher is the apparent field capacity, the slower is its attainment, and the less distinct is its value.

(2) Type of clay present: The higher the content of montmorillonite is, the greater is the content of water adsorbed and retained at any time.

(3) Organic matter content: Soil organic matter can help retain more water, though the amount of organic matter normally present in mineral soils is too low to have much of an effect.

(4) Depth of wetting and antecedent moisture: In general (but not always), the wetter the profile is at the outset, and the greater is the depth of wetting during infiltration, the slower is the rate of redistribution, and the greater is the apparent field capacity.

(5) The presence of impeding layers in the profile, such as layers of clay, sand, or gravel, can inhibit redistribution and increase the observable field capacity.

(6) Evapotranspiration: The rate and pattern of the upward extraction of water from the soil can affect the gradients and flow directions in the profile and thus modify the redistribution process.

G. Summary

In the absence of groundwater and provided the soil profile is sufficiently deep, the typical moisture profile at the end of the infiltration process consists of a wetted zone in the upper part of the profile, and a relatively dry zone beneath. The post-infiltration movement of water from the moister to the drier zones within the profile is called redistribution. This process can be caused by matric suction gradients, or gravity, or both, and it is affected by hysteresis. The rate of redistribution generally decreases with time, as the suction gradients decrease and the hydraulic conductivity of the desorbing zone falls off. Thus, moisture loss from the upper zone, rapid at first, becomes slower and slower, and in time this loss can become imperceptible. The seemingly stable wetness remaining has long been termed "field capacity," and taken to represent the upper limit of moisture availability in the field. The field capacity is not an equilibrium value or a true constant, since redistribution does not generally cease within a few days and can persist for very long periods of time. In any case, different soils vary greatly, sometimes by a factor of ten, in their ability to retain water after comparable periods following infiltration. Soil wetness at any given time after infiltration depends on the soil's hydraulic properties and profile uniformity (or layering sequence), as well as on the quantity of water infiltrated and the initial (pre-infiltration) wetness.

8 *Groundwater Drainage*

A. General

As described in the preceding chapter, the term redistribution pertains to the movement of water in a predominantly unsaturated soil. The term "groundwater drainage," on the other hand, normally applies to the movement of water in the saturated zone, and, more specifically, to the outflow or artificial removal of excess water from the soil, generally by lowering the water table[1] (or by preventing its rise).

Soil saturation as such is not necessarily harmful to plants. The roots of most plants can, in fact, thrive in water, provided it is free of toxic substances and contains sufficient oxygen to allow normal respiration. As is well known, plant roots must respire constantly, since most plants are unable to transfer the required oxygen from their canopies to their roots. Excess water can block soil pores, and thus retard aeration and effectively strangulate the roots. In water-logged soils, gas exchange with the atmosphere is restricted to the surface few inches of the soil, while within the profile proper, oxygen may be totally lacking, and carbon dioxide may accumulate to harmful levels. Under anaerobic conditions, various substances are reduced from their normally oxidized states. Toxic concentrations of ferrous, sulfide, and manganous ions can develop. These, in combination with products of the anaerobic decomposition of organic matter (e.g., methane) can greatly inhibit plant growth. At the same time, nitrification is prevented, and various plant and root diseases (especially fungal) are more prevalent.

The occurrence of a high water table condition may not always be evident

---

[1] The removal of free water tending to accumulate over the soil surface is often termed "surface drainage" and is outside the scope of our discussion.

at the surface, which may appear to be dry even while the soil is completely water-logged beyond a depth of one or two feet. Where the effective rooting depth is thus restricted, plants may suffer not only from lack of oxygen in the soil, but also from lack of nutrients, and, where the water table drops periodically, from lack of water as well. Excessive moisture causes the soil to be susceptible to compaction by animal and machinery traffic. Furthermore, a wet soil does not warm up readily in the spring, and may thus retard germination and early plant growth.

Temperature also has an effect on plant sensitivity to restricted drainage. A rise in temperature is associated with a decrease in the solubility of oxygen in water and with an increase in the respiration rate of both plant roots and soil microorganisms. The damage caused by excessive soil moisture is therefore likely to be greater in a warm climate than in a cold one. Moreover, in a warm climate, the evaporation rate and, hence, the hazard of salinity are likely to be greater than in a cool climate. The process of evaporation inevitably results in the deposition of salts at or near the soil surface, and these salts can be removed and prevented from accumulating only if the water table remains deep enough to permit leaching without subsequent resalinization through capillary rise of the groundwater. Irrigated lands, even in arid regions, frequently require drainage. In fact, irrigation without drainage can be disastrous. Once-thriving civilizations based on irrigated agriculture in river valleys (as in Mesopotamia, for instance) have been destroyed through the insidious, and for a time invisible, process of salt accumulation caused by poor drainage.

In many regions, such as coastal plains and river valleys, large tracts of land that are potentially highly productive lie waste, or are of restricted use, because of excessive moisture. Wherever topographic conditions, soil imperviousness, or the presence of shallow groundwater prevent the profile from draining itself adequately, the soil may become an unsuitable medium for plant growth unless drained artificially. Artificial drainage of such lands can result in the reclamation of millions of acres for the production of food and fiber for the world's growing population. In large areas, likewise, drainage is required for the long-term maintenance of soil productivity. Irrigated agriculture cannot long be sustained in many arid regions unless drainage is provided for salinity control as well as for effective soil aeration (Richards, 1954; Fireman, 1957).

On the other hand, the presence of a shallow water table in the soil profile (provided it is not too shallow) can in certain circumstances be beneficial. Where precipitation or irrigation water is scarce, the availability of groundwater within reach of the roots can supplement the water requirements of crops. However, to obtain any lasting benefit from the presence of a water table in the soil, its level and fluctuation must be controlled.

Numerous investigations of groundwater flow and drainage have resulted in a very extensive body of literature on this subject. Reference should be made particularly to the monograph edited by Luthin (1957).

B. Groundwater Flow Phenomena

While water in unsaturated soil is strongly affected by suction gradients and its movement is subject to very considerable variations in conductivity resulting from changes in soil wetness, groundwater is always under positive hydrostatic pressure and hence it saturates the soil. Thus, no suction gradients and no variations in wetness or conductivity normally occur below the water table[2] and the hydraulic conductivity is maximal and fairly constant in time (though it may vary in space and direction).

Despite these differences, however, the saturated and unsaturated zones are not separate realms but parts of a continuous flow system. The groundwater is often sustained and its position determined by the rate of percolation of precipitation or irrigation water through the unsaturated zone. Reciprocally, the position of the water table affects the moisture profile and flow conditions above it. One problem encountered in attempting to distinguish between the unsaturated and saturated zones is that the boundary between them may not be at the water table (defined as the locus of points at which the hydrostatic pressure is equal to atmospheric pressure), but at some elevation above it, corresponding to the extent of the capillary fringe (at which the suction is equal to the air-entry value for the soil). Frequently, this boundary is diffuse and scarcely definable, particularly when affected by hysteresis.

The various circumstances, or boundary conditions, under which groundwater movement can occur were elucidated by Kirkham (1957), van Schilfgaarde (1957), and Childs (1969).

Water may seep into or out of the groundwater through the soil surface, or by lateral flow through the sides of a channel or a porous drainage tube. In general, the groundwater table, though hardly ever entirely flat, seldom exhibits steep gradients (except in the drawdown region about drainage channels, tubes, or wells). Where the land surface elevation varies, as well as where the amount of infiltration water supply varies areally, the water table can vary in depth and may in places and at times intersect the soil surface and emerge as free (ponding) water above it.

If the depth of the water table is constant, the indication is that the rate of inflow into the groundwater and the rate of outflow are equal. For instance,

---

[2] We are disregarding here possible effects due to overburden pressures and swelling phenomena (Philip, 1969).

where there is a net downward seepage of infiltration water, this must be offset by downward or horizontal outflow of the groundwater if the water table is to remain stationary. On the other hand, a rise or fall of the water table indicates a net recharge or discharge of groundwater, respectively. Such vertical displacements of the water table can occur periodically, as under a seasonally fluctuating regimen of rainfall or irrigation. The rise and fall of the water table can also be affected by barometric pressure changes.

Geometrically, groundwater flow can be quite complicated where the profile is layered or anisotropic, or where the sources and sinks of water are distributed unevenly. If the profile above the water table consists of a sequence of layers such that a highly conductive one overlies one of low conductivity, then it is possible for the flow rate into the top layer to exceed the transmission rate through the lower layer. In such circumstances, the accumulation of water over the interlayer boundary can result, temporarily at least, in the development of a *perched* (or secondary) *water table* with positive hydrostatic pressures. If infiltration persists at a relatively high rate, the two bodies of water will eventually merge. If infiltration ceases, the perched water table will eventually tend to disappear by downward seepage into the primary water table.

In some cases, groundwaters of different density (caused by salinity or temperature differences) can come into contact. One such case, of common occurrence along sea coasts, is that of a body of fresh water (called a Ghyben–Herzberg lens) floating over a body of saline water. At the contact plane, or interface, of the two bodies of water, miscible displacement phenomena can be observed.

C. Flow Equations

Darcy's law alone is sufficient to describe only steady flow processes. In general, however, Darcy's law must be combined with the mass-conservation law, to obtain the general flow equation for homogeneous isotropic media:

$$\frac{\partial \theta}{\partial t} = K\left(\frac{\partial^2 H}{\partial x^2} + \frac{\partial^2 H}{\partial y^2} + \frac{\partial^2 H}{\partial z^2}\right) \tag{8.1}$$

where θ is wetness, t time, K conductivity and H hydraulic head. In a saturated, stable medium, there is no change of wetness (water content) with time, and we obtain:

$$K_s\left(\frac{\partial^2 H}{\partial x^2} + \frac{\partial^2 H}{\partial y^2} + \frac{\partial^2 H}{\partial z^2}\right) = 0 \tag{8.2}$$

where K_s is the saturated conductivity. Equation (8.2) is known as the Laplace equation. Since K_s is not zero, it follows that

$$\frac{\partial^2 H}{\partial x^2} + \frac{\partial^2 H}{\partial y^2} + \frac{\partial^2 H}{\partial z^2} = 0 \qquad (8.3)$$

The expression $(\partial^2/\partial x^2 + \partial^2/\partial y^2 + \partial^2/\partial z^2)$, or, in vector notation, ∇^2, is known as the Laplacian operator. Accordingly, we can write Laplace's equation as: $\nabla^2 H = 0$.

If, instead of using Cartesian coordinates (x, y, z), we cast Eq. (8.3) into cylindrical coordinates (r, α, z), we obtain

$$\frac{1}{r}\frac{\partial}{\partial r}\left(r\frac{\partial H}{\partial r}\right) + \frac{1}{r^2}\frac{\partial^2 H}{\partial \alpha^2} + \frac{\partial^2 H}{\partial z^2} = 0 \qquad (8.4)$$

Laplace's equation also applies to systems other than fluid flow in porous media, namely to the flow of heat in solids and of electricity in electrical conductors. Solutions for boundary values appropriate to the latter systems, some of which are also applicable to soil-water flow, are given in the books by Smythe (1950) and by Carslaw and Jaeger (1959).

The direct analytical solution of Laplace's equation for conditions pertinent to groundwater flow is not generally possible. Therefore, it is often necessary to resort to approximate or indirect methods of analysis. Where flow is restricted to two dimensions, the equation becomes

$$\frac{\partial^2 H}{\partial x^2} + \frac{\partial^2 H}{\partial y^2} = 0 \qquad (8.5)$$

which is more readily soluble (Childs, 1969).

A well-known numerical technique is the method of relaxation (Southwell, 1946), in which the potentials and flow rates throughout the system are calculated successively and reiteratively in reference to known boundary values. Newer techniques are those of successive overrelaxation (Smith, 1965), and alternating-direction implicit methods (Peaceman and Rachford, 1955; Varga, 1962).

Approximate solutions of drainage flow problems were reviewed by van Schilfgaarde (1957), who stressed that the simplifications possible in a less than rigorous approach can be of considerable value in practice, but that they require a constant awareness of the limitations which are inherent in the use of such simplifications. His discussion encompasses both steady-state processes (in which the potentials in the system do not change with time) and nonsteady, or transient-state, processes (during which the potentials and fluxes vary).

In the solution of problems relating to flow toward a shallow sink (a drainage tube or ditch), it is often convenient to employ the Dupuit–

Forchheimer assumptions (Forchheimer, 1930) that, in a system of gravity flow toward a shallow sink, all the flow is horizontal and that the velocity at each point is proportional to the slope of the water table but independent of depth. Though these assumptions are obviously not correct in the strict sense and can in some cases lead to anomalous results (Muskat, 1946), they often provide feasible solutions in a form simpler than obtainable by rigorous analysis. They are most suitable where the flow region is of large horizontal extent relative to its depth.

The equation for the free-water surface can be written as (van Schilfgaarde, 1957):

$$\frac{\partial^2 h^2}{\partial x^2} + \frac{\partial^2 h^2}{\partial y^2} = 0 \tag{8.6}$$

where h is the height of the saturated soil above an impervious stratum and x and y are the horizontal coordinates. This equation permits determination of the shape of the free-water surface and the velocity at any point for a shallow gravity-flow system in steady state.

D. Analysis of Falling Water-Table

Truly steady-state flow processes are rare. More typical are transient-state processes, which involve a change in water-table height. Problems of this sort are more difficult to solve. Drainage under such conditions has been described in terms of the *specific-yield* concept, generally defined as the volume of water extracted from the groundwater per unit area when the water table is lowered a unit distance (or as the ratio of drainage flux to rate of fall of the water table). The assumption that there exists such a distinct "drainable" porosity, and that this fraction of the soil volume drains instantaneously as the water table descends, may be seriously misleading. In actual fact, the volume of water drained increases gradually with increasing suction which accompanies the gradual descent of the water table.

According to Luthin (1966), the drainable porosity f_d is a function of the capillary pressure (i.e., the negative pressure head, or the suction head), h, and can be written as $f_d(h)$. As the water table drops from h_1 to h_2, the quantity of water drainage out of a unit column will be

$$Q = \int_{h_2}^{h_1} f_d(h) \, dh \tag{8.7}$$

The function $f_d(h)$ is related to the soil moisture-characteristic and is generally a complex function. However, it is sometimes possible to write an approximate expression for the relation of drainable porosity to capillary pressure and still end up with a reasonable prediction of the amount of water that drains out of a profile of soil in which the water table is dropping. The simplest equation to use is that of a straight line $f_d = ah$, where a is the slope of the line. The quantity of water drained is now given by

$$Q = \int_{h_2}^{h_1} ah\, dh = \frac{a}{2}(h_1^2 - h_2^2) \qquad (8.8)$$

Problems involving falling water tables have been handled by representing the transient process as a succession of steady states (Childs, 1947; Kirkham and Gaskell, 1951; Collis-George and Youngs, 1958; Isherwood, 1959; Bouwer and van Schilfgaarde, 1963). Childs and Poulovassilis (1962) showed that the soil-moisture profile above a descending water table depends upon the time-rate of that descent.

The time-dependent vertical drainage of a soil column following a drop of the water table has been studied by Day and Luthin (1956), Youngs (1960), Gardner (1962), Jensen and Hanks (1967), and Jackson and Whisler (1970).

Gardner assumed a constant mean diffusivity and a linear relation between hydraulic head and soil wetness. His equation is

$$\frac{Q}{Q_\infty} = 1 - \frac{8}{\pi^2} \exp\left(-\frac{\bar{D}\pi^2 t}{4L^2}\right) \qquad (8.9)$$

where Q is the volume of water per unit area removed during time t, Q_∞ is the total drainage after infinite time, $\bar{D}$ is the weighted mean diffusivity, and L is the column length.

Youngs' derivation was based on the capillary-tube model and on the assumption of a distinct *drainage front*, i.e., that a sharp demarcation of constant suction exists between the saturated and the drained zones. This implies that an equal quantity of water drains for an equal downward advance of the drainage front. The above assumption is analogous to the Green and Ampt assumption pertaining to infiltration (namely, that there exists a distinct wetting front which remains at constant suction as it advances down the profile). The approximate equation obtained is

$$\frac{Q}{Q_\infty} = 1 - \exp\left(\frac{q_0 t}{Q_\infty}\right) \qquad (8.10)$$

where q_0 is the initial flux of drainage at the base of the column (the water table) and the other variables are as above.

Youngs' approach was modified by Jackson and Whisler, who assumed that hydraulic conductivity is linearly related to average soil wetness during drainage and that the latter is related to the average length of soil column yet to be drained. They also assumed a constant drainable porosity. They claimed that their theory accords with experimental data over a much longer time than previously published theories.

In general, the applicability of such approximate theories depends on the validity of the assumptions made in their derivation. Some approximate solutions which provide reasonably accurate predictions of drainage outflow do not give an accurate description of the changing moisture and suction profiles above the water table. More exact solutions of nonsteady drainage problems can be obtained by numerical techniques (e.g., Jensen and Hanks, 1967), but such procedures are still too tedious and time-consuming for routine use, requiring the use of a computer for each case considered.

Most of the equations available for the estimation of drainage from soils disregard the possible effect of evapotranspiration. If evapotranspiration occurs while the soil is draining, the amount of drainage will obviously be lessened, and equations such as those cited will give overestimations.

E. Flow Nets, Models, and Analogs

One interesting (though often laborious) method for describing ground-water flow is the use of a *flow net*, which is a graphical representation of the distribution of hydraulic potentials, gradients, flow directions, and fluxes within the soil. The flow net includes lines, or curves, of two types: *equipotentials*, which are curves joining points of equal hydraulic head, and

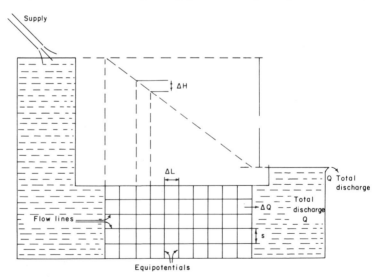

Fig. 8.1. Flow net in a constant-head permeameter.

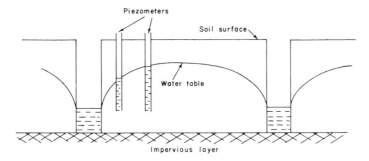

Fig. 8.2. Piezometers in a system of groundwater drainage by means of open ditches.

streamlines, which are perpendicular to the equipotentials and represent the direction of flow from higher to lower potential. The space between each pair of adjacent equipotentials represents a constant potential drop, while the space between each pair of adjacent streamlines represents a constant discharge. When the complete flow net is known for any system, and the hydraulic conductivity is also known, it is possible to calculate the exact fluxes and their directions in different parts of the soil.

The simplest example of a flow net is the one describing a constant-head permeameter, as shown in Fig. 8.1, in which water flows through a soil model of unit thickness (in the plane perpendicular to the page) of height S and of length L. In this simple case, all of the flow lines are straight and parallel. The number of flow lines drawn in the net is arbitrary, but their spacing is such that the discharge rates between each adjacent pair are equal.

In the simple case shown, there is actually no need to draw a flow net, since the geometry is one-dimensional and one can use Darcy's law simply and directly. However, in most cases in the field, the lines are neither straight nor parallel. In any case, however, flow nets are so drawn that the streamlines intersect the equipotentials perpendicularly,[3] and that the discharge rate Q in the section bounded by each pair of streamlines is equal to the total discharge divided by the number of such sections in the flow net. Similarly, the head drop in the sections between pairs of equipotentials ΔH is also constant.

In the field, flow conditions can be determined by the measurement of hydraulic head at different points in the soil. This is generally done by means of *piezometers*, which are borings made to below the water table, to allow repeated measurements of the hydraulic head at various sites (Fig. 8.2).

[3] Strictly speaking, this principle holds only in an isotropic soil. A nonisotropic soil (one with directionally variable conductivity, such as might be caused by a particular pattern of fissures or layers) may exhibit streamlines which are not orthogonal to the equipotential surfaces. In such cases, the flow direction is not necessarily that of the steepest potential gradient.

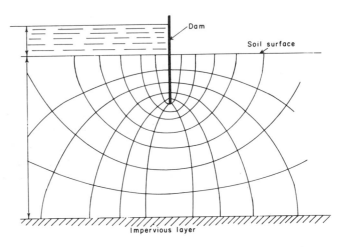

Fig. 8.3. Hypothetical flow net for seepage out of reservoir.

Hypothetical flow nets for different systems are given in Figs. 8.3 and 8.4. Where the streamlines converge, as they do close to the drains, the flux is greatest.

In a layered soil, the streamlines are often refracted in passage from one layer to the next. This has the effect of changing the flow direction in all cases except where this direction is perpendicular to the interlayer boundary (or where the two layers happen to have equal conductivities). The following relationship holds (Casagrande, 1937):

$$K_1 \cot \alpha = K_2 \cot \beta \qquad (8.11)$$

where K_1 and K_2 are the conductivities of the two adjacent layers, and α and

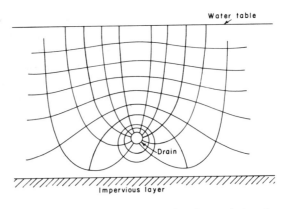

Fig. 8.4. Hypothetical flow net for flow into a drain tube.

β are the incident and emergence angles, respectively, of the streamlines as they pass from layer 1 to layer 2.

Hydraulic models are often used to help visualize drainage systems. Such models generally consist of containers of various dimensions filled with uniform or layered soil, in which the boundary and flow conditions of a given flow problem are simulated. The side wall of such a container is generally made of transparent glass or plastic, allowing visual observation of flow. Dyes are sometimes injected at various points to help in tracing the streamlines. One hydraulic model that has often been used in groundwater hydrology is the *Hele–Shaw model*, in which a fluid or fluids (generally oils of higher viscosity than water) are caused to flow between parallel glass plates. The effective hydraulic conductivity at various sections of the system can be varied by changing the separation width between the plates.

Electrical analogs are also used widely in simulating groundwater flow. Their applicability is based on the formal similarity between Ohm's and Darcy's laws[4]. Numerous types of electrical analogs have been proposed to aid in the solution of drainage and soil-water flow problems, from simple conducting sheets to complex arrays of variable resistors and capacitors (e.g., Bouwer, 1962, 1964).

F. Factors Influencing Drainage

Groundwater drainage is generally carried out by means of ditches, pipes, or " mole channels," into and within which the groundwater flows as the result of the hydraulic gradients existing in the soil. The flow rate is affected by the following factors:

[4] Ohm's law states the linear relation of electric current to the electric potential gradient. It is usually written as:

$$I = E/R$$

wherein I is current in amperes (a flow rate of one coulomb per second), E is the potential drop (volts) and R is resistance (ohms). Ohm's law can also be written in terms of conductivity, K_e. Since the total resistance of a conductor varies directly as the length L, and inversely as the area A, and the conductivity K_e, we can write

$$I = AK_e E/L$$

or

$$i = AK_e E/L$$

where i is the flow rate per unit area (the flux) and E/L is the potential gradient. This form is analogous to the way Darcy's law is usually written, namely

$$q = K\Delta H/L$$

(1) The hydraulic conductivity of the soil, which may differ from one layer or place to another if the soil is heterogeneous and may also vary directionally if the soil is anisotropic.

(2) The configuration of the water table and the hydraulic pressure of the groundwater. The water table is not always horizontal or of constant depth. Furthermore, in some cases, the groundwater may be confined and exhibit artesian pressure.

(3) The depths of the channels or tube drains, relative to the groundwater table and to the soil surface, as well as the slopes of these drains and their outlet elevation.

(4) The inlet openings along the drains. It is common practice to leave small gaps between the sections of the drainage tubes in order to allow inflow from the soil. In some cases, the drainage tubes are embedded in gravel to increase the discharge and to prevent clogging of the inlet openings by soil scouring or collapse.

(5) The horizontal spacing between the parallel drains.

(6) The diameters of the drains.

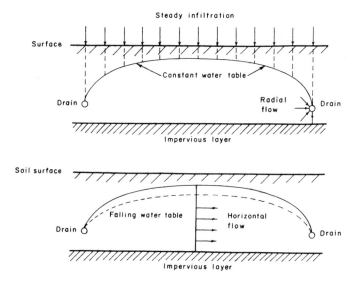

Fig. 8.5. Groundwater drainage under steady flow conditions (infiltration rate equals drainage rate and the water table remains at constant depth), and under unsteady flow resulting in a falling water table.

(7) The rate at which water is added to the groundwater by infiltration, by internal redistribution, or by lateral flow from an external source of water (i.e., outside the field being drained). A steady flow condition, in which the rate of inflow by infiltration is equal to the rate of outflow by drainage, is depicted in Fig. 8.5, in comparison with an unsteady flow condition resulting in a falling water table.

G. Drainage Design Equations

Various equations, empirically or theoretically based, have been proposed for the purpose of determining the desirable depths and spacings of drain pipes or ditches in different soil and ground-water conditions. Since field conditions are often complex and highly variable, these equations are generally based upon assumptions which idealize and simplify the flow system. The available equations are therefore approximations which should not be applied blindly. Rather, the assumptions must be examined in the light of all information obtainable concerning the circumstances at hand.

One of the most widely applied equations is that of Hooghoudt (1937), designed to predict the height of the water table which will prevail under a given rainfall or irrigation regime when the conductivity of the soil and the depth and horizontal spacing of the drains are known. This equation, like others of its type, oversimplifies the real field situation, as it disregards additional factors which may have a bearing upon groundwater movement, such as the rate of evapotranspiration, soil layering, etc. It is based on the following tacit assumptions: (1) the soil is homogeneous and of constant hydraulic conductivity; (2) the drains are parallel and equally spaced; (3) the hydraulic gradient at each point beneath the water table is equal to the slope of the water table above that point (this gradient is generally directed toward the nearest drain); (4) Darcy's law applies; (5) an impervious layer exists at a depth below the drain; (6) the supply of water from above, due to rain or irrigation, is at a constant flux q. The shape of the water table between parallel drains is generally described as elliptical.

To derive the Hooghoudt equation, let us examine flow in a profile section of a field, having a thickness of one unit, and bounded at its sides by two adjacent drains (tubes or ditches) a distance S apart. Assuming symmetry, we can draw a vertical midplane between the drains which will divide flow toward one plane from flow to the other. Now let us consider flow toward one of the drains through any arbitrary vertical plane located a distance x from that drain. The quantity of water passing through this plane per unit time must be equal to the infiltration flux multiplied by the width from the arbitrary plane to the midplane between the drains. This width is $S/2 - x$. Accordingly, the horizontal flow per unit time through the arbitrary plane is

$$Q = q\left(\frac{S}{2} - x\right) \tag{8.12}$$

At the same time, Q can be obtained from Darcy's law. If we assume the effective gradient to be equal to the slope of the water table (dh/dx) at the arbitrary vertical plane, we get

$$Q = Kh\frac{dh}{dx} \tag{8.13}$$

where K is the hydraulic conductivity and h is the height of the water table above an impervious layer which is assumed to form the "floor" of the flow system. Now we can equate the two equations

$$q\left(\frac{S}{2} - x\right) = Kh\frac{dh}{dx} \tag{8.14}$$

or

$$\tfrac{1}{2}\,qS\,dx - qx\,dx = Kh\,dh$$

which can be integrated to yield

$$qSx - qx^2 = Kh^2 \tag{8.15}$$

Assuming that at $x = 0$ (i.e., at the drain) $h = d$ (the height of the drain above the impervious floor), while at $x = S/2$ (the midplane) $h = H + d$ (where H is the maximal height of the water table above the drains) we obtain Hooghoudt's equation

$$S^2 = \frac{4KH}{q}(2d + H) \tag{8.16}$$

This equation has been used widely for determining the desirable spacing and depth of drains needed to maintain the water table below a certain level. That level, as well as the average infiltration flux and hydraulic conductivity, must be known *a priori*. A depth must also be known, or assumed, for the impervious layer.

Among the most serious weaknesses of this approach are the assumptions of an impervious layer at some definable shallow depth and the disregard of that portion of the total flow which occurs above the water table (Donnan, 1947; Bouwer, 1959).

Other equations, derived by more rigorous procedures, have been offered by Kirkham (1958) and by the U.S. Bureau of Reclamation (Luthin, 1966).

The ranges of depth and spacing generally used for the placement of drains in field practice are shown in Table 8.1.

Table 8.1

PREVALENT DEPTHS AND SPACINGS OF DRAINAGE TUBES IN VARIOUS SOIL TYPES

Soil type	Hydraulic conductivity (cm/day)	Spacing of drains (m)	Depth of drains (m)
Clay	0.15	10–20	1–1.5
Clay loam	0.15–0.5	15–25	1–1.5
Loam	0.5–2.0	20–35	1–1.5
Fine, sandy loam	2.0–6.5	30–40	1–1.5
Sandy loam	6.5–12.5	30–70	1–2
Peat	12.5–25	30–100	1–2

H. Summary

The need for artificial drainage arises whenever excessive amounts of water in the rooting zone limit aeration or prevent removal of excess salts. Such conditions are not always visible at the soil surface, and the problem often requires detailed investigation before an efficient drainage system can be devised. The problem is likely to be most acute in irrigated areas and in fine-textured soil, where irrigation without drainage can cause the water table to rise unduly. Planning drainage systems involves knowledge of water-table heights, water supply rates, and the conductive properties of the soil.

9 *Evaporation from Bare-Surface Soils*

A. General

Evaporation in the field can take place from plant canopies, from the soil surface, or, more rarely, from a free-water surface. Evaporation from plants, called *transpiration*, is the principal mechanism of soil-water transfer to the atmosphere when the soil surface is covered with vegetation. Soil-water uptake and transpiration by plants is, however, the subject of the next chapter.

When the surface is at least partially bare, evaporation can take place from the soil as well as from plants. Since it is generally difficult to separate these two processes, they are commonly lumped together and treated as a single process called *evapotranspiration*.

In the absence of vegetation, and when the soil surface is subject to radiation and wind effects, evaporation occurs directly and entirely from the soil. This process is the subject of the present chapter. It is a process which, if uncontrolled, can involve very considerable losses of water in both irrigated and unirrigated agriculture. Under annual field crops, the soil surface may remain largely bare throughout the periods of tillage, planting, germination, and early seedling growth, periods in which evaporation can deplete the moisture of the surface soil and thus affect the growth of young plants during their most vulnerable stage. Rapid drying of a seedbed can thwart germination and thus doom an entire crop from the start. The problem can also be acute in young orchards, where the soil surface is often kept bare continuously for several years, and in dryland farming in arid zones, where the land is regularly fallowed for several months to collect and conserve rainwater from one season to the next.

Evaporation of soil water involves not only loss of water but also the danger of soil salinization. This danger is felt most in regions where irrigation water is scarce and possibly brackish and where annual rainfall is low, as well as in regions with a high groundwater table.

B. Physical Conditions

Three conditions are necessary if the evaporation process from a given body is to persist. Firstly, there must be a continual supply of heat to meet the latent heat requirement (which is about 590 cal/gm of water evaporated at 15°C). Secondly, the vapor pressure in the atmosphere over the evaporating body must remain lower than the vapor pressure at the surface of that body[1] (i.e., there must be a vapor-pressure gradient between the body and the atmosphere), and the vapor must be transported away, either by diffusion or convection, or both. These two conditions—namely, supply of energy and removal of vapor—are external to the evaporating body and are influenced by meteorological factors such as air temperature, humidity, wind velocity, and radiation, which together determine the *atmospheric evaporativity* (being the maximal flux at which the atmosphere can vaporize water from a free-water surface).[2]

The third condition is that there be a continual supply of water from or through the interior of the body to the site of evaporation. This condition depends upon the content and potential of water in the body as well as upon its conductive properties, which together determine the maximal rate at which the body can transmit water to the evaporation site. Accordingly, the actual evaporation rate is determined either by external evaporativity or by the soil's own ability to deliver water, whichever is the lesser (and hence the limiting factor).

If the top layer of soil is initially quite wet, as it typically is at the end of an infiltration episode, the process of evaporation into the atmosphere will generally reduce soil wetness and thus increase matric suction at the surface. This, in turn, will generally cause soil water to be drawn upward from the layers below, provided they are sufficiently moist.

---

[1] Even in a humid region, the atmosphere on a clear day is likely to be at a relative humidity equivalent to a negative water potential of many score or even hundreds of bars.

[2] Atmospheric evaporativity, also called "the evaporative demand of the atmosphere," is not entirely independent of the properties of the evaporating surface. For instance, the net supply of energy for evaporation is affected by the reflectivity, emissivity, and thermal conductivity of the surface soil. Hence, the evaporative demand acting on a soil will not be exactly equal to evaporation from a free-water surface. The latter itself depends on the size and depth of the water body considered.

Among the various sets of conditions under which evaporation may occur are the following:

(1) A shallow groundwater table may be present at a constant or variable depth—or it may be absent (or too deep to affect evaporation). Where a groundwater table occurs close to the surface, steady-state flow may take place from the saturated zone beneath, through the unsaturated layer to the surface. Continued evaporation can thus occur without changing the soil-moisture content (though cumulative salinization may take place at the surface). In the absence of shallow groundwater, on the other hand, the loss of water at the surface and the resulting upward flow of water in the profile will necessarily be a transient-state process causing the soil to dry.

(2) The soil profile may be uniform (homogeneous and isotropic), or its properties may change gradually in various directions, or the profile may consist of distinct layers differing in texture or structure.

(3) The profile may be shallow ("finite-depth") or deep ("semi-infinite"). It may also be effectively semiinfinite for a time, then become finite as the effect of the process begins to be felt at the bottom boundary.

(4) The flow pattern may be one-dimensional (vertical), or two- or three-dimensional (as in the presence of vertical cracks which form secondary evaporation planes inside the profile).

(5) Conditions may be isothermal or nonisothermal. In case of the latter, the thermal gradients and conduction of heat and vapor through the system may interact with liquid water flow.

(6) External environmental conditions may remain constant or fluctuate.

(7) Soil moisture flow may be governed by evaporation alone, or by both evaporation (at the top of the profile) and internal drainage, or re-distribution, down below.

C. Capillary Rise from a Water Table

The rise of water in the soil from a free-water surface (i.e., a water table) has been termed *capillary rise*. This term derives from the "capillary model," which regards the soil as analogous to a bundle of capillary tubes, predominantly wide in the case of a sandy soil and narrow in the case of a clay soil. Accordingly, the equation relating the equilibrium height of capillary rise h_c to the radii of the pores is

$$h_c = \frac{2\gamma \cos \alpha}{r\rho_w g} \tag{9.1}$$

where γ is the surface tension, r the capillary radius, ρ_w the water density, g the gravitational acceleration, and α the wetting angle, normally taken as zero. This equation predicts that water will rise higher (though less rapidly) in a clay than in a sand. However, soil pores are not capillary tubes of uniform or constant radii, hence the height of capillary rise will differ in different pores. Above the water table, matric suction will generally increase with height and the number of water-filled pores will decrease in each soil.

The rate of capillary rise generally decreases with time as the soil is wetted to greater height and as equilibrium is approached.

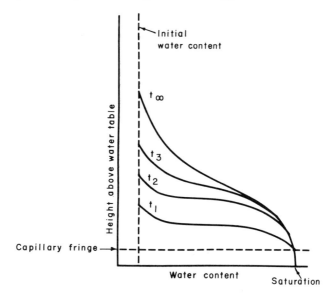

Fig. 9.1. The upward infiltration of water from a water table into a dry soil: water-content distribution curves ("moisture profiles") for various times ($t_1 < t_2 < t_3 < t_\infty$). t_∞ is the profile after infinitely long time (equilibrium).

The wetting of an initially dry soil by upward capillary flow, illustrated in Fig. 9.1, is a rare occurrence in the field. In its initial stages, this process is similar to infiltration, except that it takes place in the opposite direction. At later stages of the process, the flux does not tend to a constant value, as in downward infiltration, but to zero. The reason is that the direction of the gravitational gradient is opposite to the direction of the matric suction gradient, and when the latter (which is large at first but decreases with time) approaches the magnitude of the former, the overall hydraulic gradient approaches zero.

Such an ideal state of static equilibrium between the gravitational head and the suction head is the exception rather than the rule under field conditions. In general, the condition of soil water is not static but dynamic— that is, soil water is constantly flowing. In the case where a water table is present, soil water does not attain equilibrium or rest even in the absence of vegetation, since the soil surface is subject to the evaporating action of the ambient atmosphere. If soil and external conditions are constant, that is, if the soil is of stable structure, the water table is at constant depth, and the atmospheric evaporativity also remains constant (at least approximately)— then, in time, a steady-state flow situation can develop from water table to atmosphere via the soil.[3]

D. Steady Evaporation in the Presence of a Water Table

The steady-state upward flow of water from a water table through the soil profile to an evaporation zone at the soil surface was first studied by Moore (1939). Theoretical solutions of the flow equation for this process were given by several workers, including Philip (1957) and Gardner (1958).
The equation describing steady upward flow is

$$q = K\left(\frac{d\psi}{dz} - 1\right) \tag{9.2}$$

or

$$q = D\frac{d\theta}{dz} - K \tag{9.3}$$

where q is flux, ψ suction head, K hydraulic conductivity, D diffusivity, and z height above the water table. The equation shows that flow stops ($q = 0$) when $d\psi/dz = 1$. Another form of Eq. (9.2) is

$$\frac{q}{K} + 1 = \frac{d\psi}{dz} \tag{9.4}$$

Integration gives

$$z = \int \frac{d\psi}{1 + q/K} = \int \frac{K}{K + q}\, d\psi \tag{9.5}$$

or

$$z = \int \frac{D(\theta)}{K(\theta) + q}\, d\theta \tag{9.6}$$

[3] Strictly speaking, evaporation in the presence of a water table will seldom, if ever, occur under truly steady-state conditions. The representation of this process as a steady-state flow is, however, a useful approximation from the analytical point of view.

In order to integrate the right-hand side of Eq. (9.5), the functional relation between ψ and K [i.e., $K(\psi)$] must be known. An empirical equation given by Gardner (1958) is

$$K = a(\psi^n + b)^{-1} \tag{9.7}$$

where parameters a, b, and n are constants which must be determined for each soil. Accordingly, Eq. (9.2) becomes

$$q = \frac{a}{\psi^n + b}\left(\frac{d\psi}{dz} - 1\right) \tag{9.8}$$

With Eq. (9.7), Eq. (9.5) can be integrated to obtain suction distributions with height for different fluxes, as well as fluxes for different surface-suction values. The theoretical solution is shown graphically in Fig. 9.2 for a fine sandy loam soil with an n value of three.

The curves show that the steady rate of capillary rise and evaporation depend on the depth of the water table and on the suction at the soil surface.

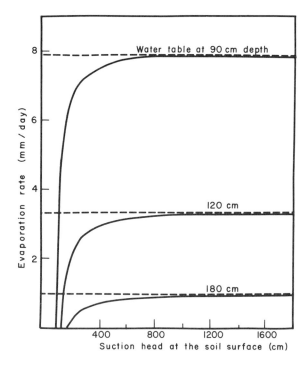

Fig. 9.2. Steady rate of upward flow and evaporation from a water table as function of the suction prevailing at the soil surface. The soil is a fine sandy loam, with $n = 3$. (After Gardner, 1958.)

This suction is dictated largely by the external conditions, since the greater the atmospheric evaporativity, the greater the suction at the soil surface upon which the atmosphere is acting. However, increasing the suction at the soil surface can increase the flux through the soil only up to an asymptotic maximal rate which depends on the depth of the water table. Even the driest and most evaporative atmosphere cannot steadily extract water from the surface any faster than the soil profile can transmit from the water table to that surface. The fact that the soil profile can limit the rate of evaporation is a remarkable and useful feature of the unsaturated flow system. The maximal transmitting ability of the profile depends on the hydraulic conductivity of the soil in relation to the suction.

Disregarding the constant b of Eq. (9.7), Gardner (1958) obtained the function

$$q_{max} = \frac{Aa}{d^n} \tag{9.9}$$

where d is the depth of the water table below the soil surface; a and n are constants from Eq. (9.7); A is a constant which depends on n; and q_{max} is the limiting (maximal) rate at which the soil can transmit water from the water table to the evaporation zone at the surface.

We can now see how the actual steady evaporation rate is determined either by the external evaporativity or by the water-transmitting properties of the soil, depending on which of the two is lower, and therefore limiting. Where the water table is near the surface, the suction at the soil surface is low and the evaporation rate is determined by external conditions. However, as the water table becomes deeper and the suction at the soil surface increases, the evaporation rate approaches a limiting value regardless of how high external evaporativity may be.

Equation (9.9) suggests that the maximal evaporation rate decreases with water-table depth more steeply in coarse-textured soils (in which the n value is greater[4]) than in fine-textured soils. Nevertheless, a sandy loam soil can still evaporate water at an appreciable rate (as shown in Fig. 9.2) even when the water table is as deep as 180 cm. Figure 9.3 illustrates the effect of texture on the limiting evaporation rate.

The subsequent contributions of a number of workers (Visser, 1959; Wind, 1959; Talsma, 1963) generally accorded with the above theory.[5]

---

[4] That is, because in sandy soils the conductivity generally falls off more steeply with increasing suction than in a clayey soil.

[5] Hadas and Hillel (1968), however, found that experimental soil columns deviated from predicted behavior, apparently owing to spontaneous changes of soil properties, particularly at the surface.

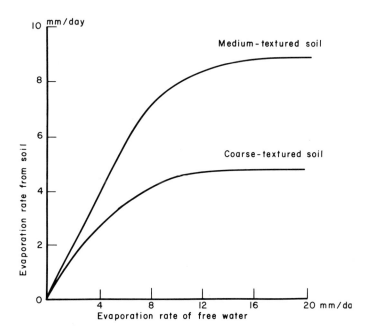

Fig. 9.3. Theoretical relation between the rate of evaporation from coarse- and medium-textured soils (water table depth, 60 cm) and the rate of evaporation from a free-water surface (after Gardner, 1958).

E. Hazard of Salinization Due to High Water Table

The rise of water from a shallow water table can in some cases serve the useful purpose of supplying water to the root zone of crops. However, this process also entails the hazard of salinization, especially where the ground-waters are brackish and potential evaporativity is high. In fine-textured soils, the danger of salinization can be appreciable even where the water table is several meters deep. The tendency for water to flow from the water table toward the soil surface will persist as long as the suction head prevailing at the surface is greater than the depth of the water table. The gradual and irreversible salinization of the soil may have been the process responsible for the destruction of once-thriving agricultural civilizations based on the irrigation of river valleys with high-water-table conditions. Excessive irrigation tends to raise the water table and thus aggravate the salinization problem.

Lowering the water table by drainage can decisively reduce the rate of capillary rise and evaporation. Drainage is a costly operation, however, and it is therefore necessary, ahead of time, to determine the optimal depth to which the water table should be lowered. Among the important considerations in this regard is the necessity to limit the rate of capillary rise to the surface (Gardner, 1958). In the soil described by Fig. 9.2, for example, the maximal rate of profile transmission to the surface is 8 mm/day when the water table is at a depth of 90 cm. Since potential evaporativity is seldom greater than this, it follows that a water-table rise above the depth of 90 cm would not be likely to increase the evaporation rate. On the other hand, a lowering of the water table to a depth of 180 cm can decrease the evaporation rate to 1 mm/day.[6] An additional lowering of the water table to a depth of 360 cm will reduce the maximal evaporation rate to 0.12 mm/day, while any further lowering of the water table can cause only a negligible reduction of evaporation and might in any case be prohibitively expensive to carry out.

Uniform profiles occur only rarely. Often, the soil consists of a number of more-or-less distinct layers. Layered conditions may in fact exist even when the soil is texturally uniform, owing to structural differentiation. Holmes *et al.* (1960) pointed out the possible effects of surface layers with different tilths upon evaporation in the presence of a water table. Talsma (1963) found that tillage reduced the evaporation rate by half and decreased suction below the tilled zone. He attributed these effects to the formation of layered conditions and to the resulting shift in the actual evaporation zone. Gardner (1958) and Gardner and Fireman (1958) showed that a dry layer at the surface reduces the steady evaporation rate in hyperbolic relation to its thickness. Water movement through such a layer occurs mainly by vapor diffusion.

A theoretical analysis of steady evaporation from two-layer profiles was given by Willis (1960). His theory was based on the assumptions that the suction at the soil surface is infinitely high (i.e., external evaporativity is infinite), and that suction is continuous at the interlayer boundary (i.e., the suctions existing on both sides of this boundary are the same). Willis proposed a graphic technique to obtain the maximal possible flux for any given two-layer profile. A numerical solution can also be obtained.

F. Evaporation in the Absence of a Water Table (Drying)

Steady evaporation from soils is not a typical occurrence, since high-water-table conditions exist only in limited areas and water table depths (as well as external conditions) seldom remain constant for very long. More commonly, soil-moisture evaporation occurs under unsteady conditions and results in a net loss of water from the soil, i.e., it results in drying. The process of drying involves considerable losses of water, especially in arid regions, where

---

[6] This is a consequence of the equation $q_{max} = Aad^{-n}$. When $n = 3$, the doubling of d (from 90 to 180 cm) will reduce q_{max} by 2^3, i.e., by a factor of 8.

these losses can amount to 50% or more of total precipitation (Hide, 1954, 1958).

Under constant external conditions, the drying process occurs in two fairly distinct stages (Fisher, 1923; Pearse *et al.*, 1949): (1) an early, *constant-rate stage*, during which the evaporation rate is determined by external and soil-surface conditions (insofar as the latter modify the effective atmospheric evaporativity acting upon the soil), rather than by the conductive properties of the profile; and (2) a *falling-rate stage*, during which the evaporation proceeds at a rate ever lower than the evaporativity and the actual rate is dictated by the ability of the soil profile to deliver moisture toward the evaporation zone. These stages are illustrated in Figs. 9.4 and 9.5.

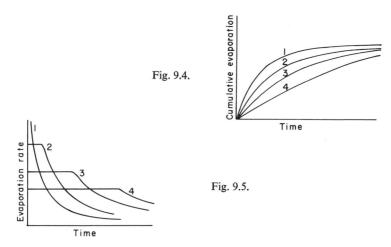

Fig. 9.4.

Fig. 9.5.

Fig. 9.4. Relation of cumulative evaporation to time (curves 1, 2, 3, 4 are in order of decreasing initial evaporation rate).

Fig. 9.5. Relation of evaporation rate (flux) to time (curves 1, 2, 3, 4 are in order of decreasing initial evaporation rate).

Any study of the drying process and of methods of controlling it must distinguish between the two stages. Since the gravitational effect is in general relatively negligible for evaporation,[7] it is possible to base an analysis on diffusivity and water-content gradient relationships. Both stages of drying depend upon the diffusivity: the first stage for its duration, the second stage for its rate.

[7] As we have already pointed out, the equivalent suction exercised by the atmosphere can amount to many (often hundreds of) bars. This suction, distributed over a few centimeters of soil depth, generally constitutes a force many times greater than gravity. For instance, a suction difference of one bar over 1 cm of soil is a head gradient of 1000, three orders of magnitude greater than the gravitational head gradient.

During the initial stage, the soil surface gradually dries out and water moves upward in response to increasing evaporation-induced gradients. The rate of evaporation can remain nearly constant as long as the increasing gradient at each point tends to compensate for the decreasing hydraulic conductivity (resulting from the decrease in water content). Sooner or later, however, the soil surface approaches equilibrium with the overlying atmosphere (i.e., becomes approximately " air-dry "). From this moment on, the suction gradients toward the surface cannot increase any more, and in fact,

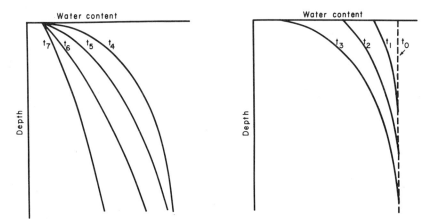

Fig. 9.6. The changing moisture profile in a drying soil. At right: the first stage of drying, during which the gradients toward the surface become steeper. At left: the second stage, in which the moisture gradients decrease as the deeper layers lose moisture by continued upward movement.

they tend to decrease as the soil in depth loses more and more moisture. Since, as the evaporation process continues, both the gradients and the conductivities at each depth near the surface are decreasing at the same time, it follows that the flux toward the surface and the evaporation rate must inevitably decrease as well. As shown in Fig. 9.5, the end of the first, i.e., the beginning of the second, stage of drying can occur rather abruptly.

This is sometimes accompanied by the downward movement into the profile of a drying front or a drying zone, so that actual evaporation may take place at some depth and the water must move through the desiccated zone by vapor diffusion. During the first stage, the soil can supply water fast enough to keep surface wetness above the value at which surface dryness begin to slow the transfer of water to the atmosphere significantly.

The gradual increase of the moisture gradients toward the surface during the first stage of drying and their gradual decrease during the second stage are illustrated in Fig. 9.6.

The length of time the initial stage of drying lasts depends upon the intensity of the meteorological factors that determine atmospheric evaporativity, as well as upon the conductive properties of the soil itself. Under similar external conditions, the first stage of drying will be sustained longer in a clayey than in a sandy soil, since clayey soils retain higher wetness and conductivity values as suction develops in the upper zone of the profile.

When external evaporativity is low, the initial, constant-rate stage of drying can persist longer. This fact has led to the hypothesis that an initially high evaporation rate may in the long run reduce cumulative moisture loss to the atmosphere. This hypothesis was raised in a number of Russian papers and cited by Lemon (1956). Gardner and Hillel (1962), on the other hand, concluded that the higher initial drying rate will in fact result in a higher cumulative loss at any time. The total water loss resulting from any finite initial evaporation rate will gradually approach (but never surpass) the total loss in the extreme case where the initial evaporation rate is infinitely high. It is at present debatable whether this pattern holds, strictly speaking, for all profile and environmental conditions or whether the presence of non-uniformities (profile layers, mulches, hysteresis effects, etc.) may modify it.

A mathematical study of the constant-rate stage of drying of both finite-length and semi-infinite soil columns was carried out by Covey (1963), who neglected gravity and used an exponential dependence of diffusivity upon wetness. For homogeneous soil columns, initially uniformly wet, Covey devised a criterion for determining when the column behaves as though it were infinitely long and when finiteness of length becomes important. When the drying rate is slow, the profile dries nearly uniformly with depth, with a relatively flat gradient of wetness.

A study of the falling-rate stage of drying was reported by Gardner (1959), who assumed infinite evaporativity (i.e., that, as the process begins, the soil surface is brought instantly to its final value of dryness). For semi-infinite soil columns, a solution of the flow equation (neglecting gravity) indicates that the cumulative evaporation E can be related linearly to the square root of time according to the equation

$$E = 2(\theta_i - \theta_0)\sqrt{Dt/\pi} \qquad (9.10)$$

The evaporative flux q, being the time derivative of E, is thus inversely proportional to the square root of time:

$$q = (\theta_i - \theta_0)\sqrt{D/\pi t} \qquad (9.11)$$

In these equations, θ_i is the initial profile wetness, θ_0 is the final (surface) wetness, and $\bar{D}$ is the weighted mean diffusivity which can be used to characterize the drying process.[8]

Using an exponential diffusivity function [namely, $D = D_0 \exp \beta(\theta - \theta_0)/(\theta_1 - \theta_0)$], Gardner obtained an approximate solution for the flow equation by means of an iterative procedure (based on Crank, 1956, p. 152). He also reported experimental results indicating that cumulative evaporation from semi-infinite columns was linear with the square root of time, as predicted by Eq. (9.10). Accordingly, the time required for soil moisture at a specified depth to fall to a given level of dryness is proportional to that depth squared. Rose (1966), using the sorptivity concept of Philip (1957), described evaporation in terms of the following equation:

$$E = st^{\frac{1}{2}} + bt \qquad (9.12)$$

where s, the sorptivity (or "desorptivity," since we are dealing with a drying process) is positive and b is negative. This approach also applies only to the falling-rate stage of drying.

The numerical procedure of Hanks and Bowers (1962), first developed for infiltration, has been applied to drying by Hanks and Gardner (1965), who studied the effects of different D vs. θ relations as well as of layering.

Gardner (1959), as well as Klute et al. (1965), studied falling-rate evaporation from finite-length columns. In this case, the flux, though initially proportional to $t^{-1/2}$, decreases more steeply as the wetness at the bottom of the column is reduced. The evaporation rate thereafter becomes roughly proportional to the product of the average diffusivity and the water content.

G. Non-Isothermal Evaporation

The discussion so far has dealt with isothermal conditions only. In recent years, it has become increasingly evident that in many cases the effect of

---

[8] Equations (9.10) and (9.11) are similar to those given for horizontal infiltration. The major difference is, however, that $\bar{D}$ must be weighted differently for drying than for wetting. According to Crank (1956), the weighted mean diffusivity for desorption processes is

$$\bar{D} = \frac{1.85}{(\theta_i - \theta_0)} 1.85 \int_{\theta_0}^{\theta_i} D(\theta)(\theta_i - \theta)^{0.85} \, d\theta$$

This weighting function yields lower values of $\bar{D}$ for corresponding θ values than the approximate weighting function for sorption processes. The need to weight the diffusivity differently can be ascribed to the fact that in infiltration the maximal flux occurs at the wet end of the column, where diffusivity is highest; whereas in drying the greatest flux is through the dry end, where diffusivity is lowest. This fact makes sorption processes inherently faster than desorption processes, and contributes to soil-moisture conservation.

temperature gradients and heat flow should not be neglected. The heat required to vaporize water must be transported to the evaporation site, which requires a nonzero temperature gradient. Nevertheless, it has generally been assumed (e.g., Miller and Klute, 1967) that the isothermal flow equation ($\partial\theta/\partial t = \partial/\partial z[D\,\partial\theta/\partial z]$) predicts many of the essential features of the constant-rate and falling-rate stages of evaporation. According to Philip (1957), the isothermal model can be expected to represent evaporation processes so long as the surface soil is not extremely dry. Recent work (Fritton, *et al.*, 1970) has indicated that the isothermal flow equation can describe cumulative evaporation quantities for both wind and radiation treatments, but that this equation cannot describe the soil water distribution pattern adequately in the case of non-isothermal flow. They concluded that where temperature gradients are important (e.g., near the soil surface) simultaneous heat and mass transfer analysis gives better fit with experimental data than the isothermal diffusion equation.[9]

Vapor movement occurs primarily in the surface zone of the soil. Where conditions are nearly isothermal, the square root of time dependence of evaporation will still hold even where much of the water movement is in the vapor phase, since isothermal vapor movement can be described by a non-linear diffusion equation of the same form as the diffusivity equation for liquid movement (Jackson, 1964).

A study of the interaction of heat and moisture flow should involve the application of the thermodynamics of irreversible processes (Wiegand and Taylor, 1961).

Fundamental studies in this field have been carried out by Taylor and Cary (1960) and by Cary (1963, 1964). Rose (1968) recently attempted to

---

[9] Simultaneous equations for analyzing non-isothermal transfer of vapor and liquid water under combined temperature and moisture gradients were offered by Philip and de Vries (1957):

$$\frac{q_w}{\rho_w} = -D_T\frac{dT}{dz} - D_\theta\frac{d\theta}{dz} - K$$

$$q_h = -\lambda\frac{dT}{dz} - \rho_w L D_{\theta\text{ vap}}\frac{dz}{d\theta}$$

where q_w is water flux density, ρ_w is density of water, D_T is thermal water diffusivity, T is temperature, z is depth, D_θ is diffusivity of soil moisture, θ is volumetric wetness, K is hydraulic conductivity, q_h is soil heat flux density, λ is thermal conductivity which includes the contribution of vapor movement, L is latent heat of evaporation, and $D_{\theta\text{ vap}}$ is vapor diffusivity (where $D_\theta = D_{\theta\text{ liq}} + D_{\theta\text{ vap}}$ and $D_{\theta\text{ liq}}$ is liquid water diffusivity). Philip (1957) eliminated the dT/dz term from the two simultaneous equations and obtained:

$$Z = -\int_{\theta(o)}^{\theta(z)}\{(\lambda D_\theta + \rho_w L D_T D_{\theta\text{ vap}})/[\lambda(K + E) - D_T g_h]\}d\theta$$

where E is evaporation rate (q_w/ρ_w in the first of the two equations above).

establish the magnitude of vapor vs. liquid water movement during evaporation under non-isothermal conditions. The effect of warming the soil is to lower the suction and raise the vapor pressure of soil water. Hence the effect of a thermal gradient is to induce flow and distilation from warmer to cooler regions. When the soil surface is warmed by radiation, this effect would tend to counter the tendency to upward flow of water in response to evaporation-induced moisture gradients. For this reason the evaporation rate might be lower when the surface is dried by radiation than when it is dried by wind of comparable evaporation potential (Hanks et al., 1967).

A combination formula derived from the energy balance and meteorological concepts was recently used by Fuchs et al. (1969) to predict hourly evaporation from a drying, bare soil.

H. Simultaneous Evaporation and Redistribution

Many of the theories and experiments concerning evaporation and drying processes relate to profiles initially wetted uniformly, usually to near-saturation. This is a convenient initial condition both from the experimental and theoretical standpoint, since, in addition to being easily reproducible, it establishes unidirectional flow and monotonic drying (uncomplicated by hysteresis), which is amenable to relatively simple analysis.

In the field, however, the process of evaporation hardly ever occurs independently of other processes. In general, the beginning of evaporation follows a wetting or irrigation, at the end of which the typical moisture profile (in the absence of a high-water-table condition) consists of a wet layer overlying relatively dry soil beneath. Under such conditions, two processes may occur simultaneously in different parts of the profile: (1) evaporation at the surface, which induces upward flow, and (2) redistribution, or internal drainage, by which water moves downward in response to gravitational and suction gradients within the deeper part of the profile. With water moving upward at the top and downward at the bottom of the wetted zone, the profile exhibits a plane of zero flux, or "watershed divide," which gradually moves downward into the profile.

Simultaneous evaporation and redistribution processes were studied by Black et al. (1969) and by Hillel et al. (1970). The two processes, though apparently occurring in different parts of the profile, may exhibit an interaction which cannot be predicted on the basis of the separate consideration of each process alone. Evaporation might reduce downward redistribution, as it might detract from the amount of water which would otherwise be available for redistribution. On the other hand, redistribution can be expected to detract from evaporation, as it tends to decrease the water content (and hence both the overall gradient and the diffusivity) in the upper zone subject to evaporation. This interaction can be important in practice, as it can enhance the effect of evaporation retardants in water conservation. Surface treatments which retard evaporation only during the initial stage of drying may have little effect on cumulative water loss in the long run in a profile in which evaporation is occurring without internal drainage. However, if such initial retardation of evaporation can allow more of the infiltrated water to move into deeper layers (beyond the reach of subsequent evaporation), surface treatments such as mulching may conserve water in the long run as well as during the initial stage.

Both evaporation and redistribution are relatively rapid at first, with the rate generally decreasing with time. Richards *et al.* (1956) and Ogata and Richards (1957) followed these processes experimentally and found that, both in redistribution and in simultaneous redistribution and evaporation, the water content W of a given zone in the profile decreased with time t according to the relation

$$W = at^{-b} \qquad (9.13)$$

or:

$$\bar{\theta} = a/Lt^{b} \qquad (9.14)$$

where a and b are empirical constants, $\bar{\theta}$ is the weighted mean wetness, and L is the depth of the zone considered.

Gardner *et al.* (1970) found experimentally that evaporation generally had little effect on the rate of redistribution. This corroborates similar statements by Rubin (1967) and by Remson *et al.* (1967), who carried out numerical analyses of redistribution processes. On the other hand, redistribution detracted greatly from evaporation. Because of the gradual decrease of wetness (and hence diffusivity) due to redistribution, cumulative evaporation can be expected to fall below the proportionality to the square root of time.

I. Reduction of Water Loss from Drying Soils

The choice of means for the reduction of evaporation depends on the stage of the process one wishes to regulate: whether it be the first stage, in which the effect of meteorological conditions on the soil surface dominates the process; or the second stage, in which the rate of water supply to the surface, determined by the transmitting properties of the profile, becomes the rate-limiting factor. Methods designed to affect the first stage cannot *a priori* be expected to serve during the second stage, and vice versa.

Covering or mulching the surface with plant residues or with reflective mulches can modify the surface energy balance and reduce the net energy available for evaporation. Thus, such surface treatments can retard evaporation during the constant-rate stage of drying (e.g., Bond and Willis, 1969). A similar effect can result from the application of materials which lower the vapor pressure of water, or from shallow cultivations designed to produce a " soil mulch " or " dust mulch " at the surface. Retardation of evaporation during the first stage can provide the plants with a greater opportunity to utilize the moisture of the uppermost soil layers, an effect which can be vital, particularly during the germination and establishment phases of plant growth. The retardation of initial evaporation can also enhance the process of re-distribution and internal drainage, and thus allow more water to migrate downward into the deeper parts of the profile, where it is conserved longer and is less likely to be lost by evaporation.

During the falling-rate stage of drying, the effect of surface treatments is likely to be only slight, and reduction of the evaporation rate and of water loss in the long run will depend on decreasing the diffusivity or conductivity of the soil profile in depth. Deep tillage, for instance, by increasing the range of variation of diffusivity with changing water content, can decrease diffusivity at the lower water contents and thus reduce the rate at which the soil can transmit water toward the surface during the second stage of the drying process.

The attempts of many farmers and field experimenters to conserve soil moisture against evaporation by surface cultivations fail all too often because, by the time the soil dries sufficiently to be cultivated, the first stage has already ended. Such cultivations, however, can have an important effect in soils which tend to form cracks. In such soils, the cracks become secondary evaporation planes, penetrating inside the soil and causing it to desiccate deeply and to an extreme degree. Timely cultivations can prevent the develop-ment of such cracks, or even obliterate cracks after they have begun to form. The quality of tillage, i.e., the degree of pulverization and clod-size distribu-tion, and the porosity of the tilled layer are also important. Coarse tillage leaves large cavities among the clods, into which air currents can penetrate and increase vapor outflow. Tillage that is too fine may pulverize the soil unduly and result eventually in a compact surface that will defeat the aim of reducing diffusivity. Each set of soil and meteorological conditions may require its own optimal depth, timing, and quality of tillage to effect the maximal reduction of evaporation losses (Hillel and Hadas, 1970).

A too-frequent irrigation regime can cause the soil surface to remain wet and the first stage of evaporation to persist most of the time, resulting in a maximal rate of water loss. The water loss by evaporation from a single deep irrigation will generally tend to be smaller than from two shallow ones with the same amount of water.

In some soils, the surface zone tends naturally to pulverize upon drying and to form a "self-mulching" layer that checks evaporation. Artificial treatments designed to hasten drying of the surface (and particularly treatments designed to "chase down" the water thermally) can in some cases reduce the eventual water loss, provided such treatments are applied early enough (Hillel, 1968).

J. Summary

In a bare soil with a shallow water table, subject to atmospheric evaporation, steady flow can take place from the groundwater source below to the evaporation sink above. When the water table is very near to the soil surface, and the soil transmits water readily, the actual evaporation rate will be limited by external evaporativity (i.e., the micrometeorological conditions). When the water table is relatively deep, the water-transmitting properties of the profile are likely to be limiting, and thus to determine the evaporation rate. Capillary rise from a water table and evaporation at the soil surface entail the hazard of progressive salinization, even though this hazard is not always immediately apparent at the surface. To avoid this hazard, which is most severe in fine textured soils under irrigation, artificial groundwater drainage may be necessary.

The rate of evaporation from a drying soil is also determined either by the external evaporativity or by the moisture-supplying and -transmitting properties of the soil profile. When the soil is relatively wet, and of relatively high conductivity, its ability to deliver water to the evaporation zone at the soil surface is not likely to be limiting, and the actual evaporation rate will tend to equal the potential rate determined by the meteorological conditions. As the soil dries, its ability to transmit and supply water is reduced and the actual evaporation rate falls below the meteorological potential rate (the atmospheric evaporativity). The best way to conserve soil moisture against evaporation is to cause it to move as deeply as possible into the profile, by proper regulation of the irrigation regimen and by controlling the initial evaporation rate so as to allow maximal time for the post-irrigation redistribution of soil water.

10 Uptake of Soil Water by Plants

A. General

Nature, despite all its laws of conservation, can in some ways be exceedingly wasteful, or so it appears at least from our own partisan viewpoint. One of the most glaring examples is the way in which it requires plants to draw quantities of water from the soil far in excess of their actual metabolic needs. In dry climates, plants growing in the field may consume hundreds of tons of water for each ton of vegetative growth. That is to say, the plants must inevitably transmit to an unquenchably thirsty atmosphere most (often considerably over 90%) of the water they extract from the soil. The loss of water vapor by plants, a process called *transpiration*, is not a necessary result of the living processes of the plants themselves. In fact, plants can thrive in a saturated or nearly-saturated atmosphere requiring very little transpiration. Rather than by plant growth, transpiration is caused by the vapor pressure gradient between the leaves and the atmosphere. In other words, it is exacted of the plants by the *evaporative demand* of the atmosphere.

In a sense, the plant in the field can be compared to the wick in an old-fashioned lamp. Such a wick has its bottom dipped into a reservoir of liquid fuel, its top subject to the burning fire which consumes the fuel, and its body constantly transmitting the liquid from bottom to top under the influence of physical suction gradients which the passive wick does not create but which are imposed upon it by the conditions prevailing at its two ends. Similarly, the plant has its roots in the soil-water reservoir and its leaves subject to the radiation of the sun and the action of the wind (i.e., the external meteorological conditions) which require it to transpire unceasingly.

This analogy, to be sure, is a gross oversimplification, since the plant is not all that passive and is in fact able at times to limit the rate of the transpiration stream by shutting the stomates of its leaves (Fig. 10.1). However,

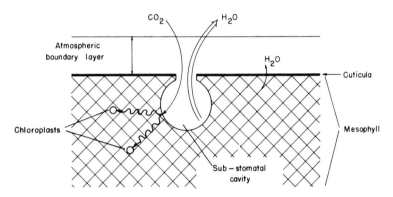

Fig. 10.1. Schematic representation of transpiration through the stomate and the cuticle, and of the diffusion of CO_2 into the stomate and to the chloroplasts (after Rose 1966).

for this limitation of transpiration the plant pays, sooner or later, in reduced growth potential, since the same stomates which transpire water also serve as foci for the uptake of the carbon dioxide needed in photosynthesis. Furthermore, reduced transpiration results in undesirable and possibly excessive warming of the plants.

To grow successfully, the plant must achieve a water economy such that the demand made upon it is balanced by the supply available to it. The problem is that the evaporative demand of the atmosphere is almost continuous, whereas rainfall occurs only occasionally and often irregularly. To survive during the dry spells between rains, the plant must rely upon the reserves of water contained in the pores of the soil.

How efficient is the soil as a water reservoir for plants? How readily can plants draw water from the soil, and to what limit can soil water continue to sustain plant growth? How is the actual rate of transpiration determined by the interaction of plant, soil, and meteorological factors? These and related questions are the topics of this chapter.

B. Classical Concepts of Soil-Water Availability to Plants

The concept of *soil-water availability*, while never clearly defined in physical terms, has for many years excited controversy among adherents of different schools of thought. Veihmeyer and Hendrickson (1927, 1949,

1950, 1955) claimed that soil water is equally available throughout a definable range of soil wetness, from an upper limit (*field capacity*) to a lower limit (the *permanent wilting point*), both of which are characteristic and constant for any given soil. They postulated that plant functions remain unaffected by any decrease in soil wetness until the permanent wilting point is reached, at which plant activity is curtailed, often abruptly. This schematized model, though based upon arbitrary limits,[1] enjoyed widespread acceptance for many years, particularly among workers in the field of irrigation management.

Other investigators, however (notably Richards and Wadleigh, 1952), produced evidence indicating that soil-water availability to plants actually decreases with decreasing soil wetness, and that a plant may suffer water stress and reduction of growth considerably before the wilting point is reached. Still others, seeking to compromise between the opposing views, attempted to divide the so-called "available range" of soil wetness into "readily available" and "decreasingly available" ranges, and searched for a "critical point" somewhere between field capacity and wilting point as an additional criterion of soil water availability.

These different hypotheses, in vogue until quite recently, are represented graphically in Fig. 10.2.

None of these schools were able to base their hypotheses upon a comprehensive theoretical framework that could take into account the array of factors likely to influence the water regime of the soil–plant–atmosphere system as a whole. Rather, they tended to draw generalized conclusions from a limited set of experiments conducted under specific and sometimes ill-defined conditions. Over the years, a great mass of empirical data has been collected, which for a long time no one knew how to explain, correlate, or resolve into a systematic theory based on physical principles.

The picture was further confused by failure to distinguish between different types of plant response to soil moisture. While transpiration rate may be, for a time, relatively independent of soil-water-content changes in the root zone, other forms of plant activity may not be. Photosynthesis, vegetative

---

[1] We have already pointed out (Chapter 7) that the field-capacity concept (Israelsen and West, 1922; Veihmeyer and Hendrickson, 1927), though useful in some cases, lacks a universal physical basis (Richards, 1960). The wilting point, if defined simply as the value of soil wetness of the root zone at the time plants wilt, is not easy to recognize, since wilting is often a temporary phenomenon, which may occur in midday even when the soil is quite wet. The *permanent wilting percentage* (Hendrickson and Veihmeyer, 1945) is based upon the *wilting-coefficient* concept of Briggs and Shantz (1912) and has been defined as the root-zone soil wetness at which the wilted plant can no longer recover turgidity even when it is placed in a saturated atmosphere for 12 hr. This is still an arbitrary criterion, since plant-water potential may not reach equilibrium with the average soil-moisture potential in such a short time. In any case, plant response depends as much on the evaporative demand (its intensity and duration) as on soil wetness.

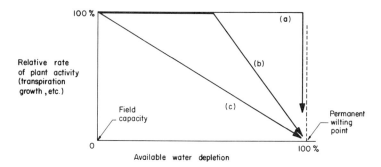

Fig. 10.2. Three classical hypotheses regarding the availability of soil water to plants: (a) equal availability from field capacity to wilting point; (b) equal availability from field capacity to a "critical moisture" beyond which availability decreases; and (c) availability decreases gradually as soil moisture content decreases.

growth, flowering, fruiting, and seed or fiber production may be related quite differently to the content or state of soil water.

It has long been recognized that soil wetness *per se* is not a satisfactory criterion for availability. Hence, attempts were made to correlate the water status of plants with the energy state of soil water, i.e., with the soil-water potential (variously termed "tension," "suction," "soil-moisture stress," etc., indeed a bewildering variety of alternative terms!). The soil-water "constants" were therefore defined in terms of potential values (e.g., $-1/10$th or $-1/3$ bar for field capacity, -15 bars for permanent wilting), which could be applied universally, rather than in terms of soil wetness (Richards and Weaver, 1944; Slater and Williams, 1965). However, even though the use of energy concepts represented a considerable advance over the earlier notions, it still fell short of taking into account the dynamic nature of soil–plant–water relations.

A fundamental experimental difficulty encountered in any attempt at an exact physical description of soil-water uptake by plants is the inherently complicated space–time relationships involved in this process. Roots grow in different directions and spacings, and as yet we have no experimental method to measure the microscopic gradients and fluxes of water in their immediate vicinity. The conventional methods for measurement of the content or potential of soil water are based on the sampling or sensing of a relatively large volume and are therefore oblivious to the microgradients toward the roots. Water suction of the soil in contact with the roots can be much greater than the average suction, as illustrated in Fig. 10.3.

An additional difficulty in describing the system physically arises from the fact that, up to the present, no satisfactory way has been found to grow plants in a soil of constant water potential. Rather, it is necessary to irrigate the soil periodically, thus refilling its effective reservoir. In a variable-soil-

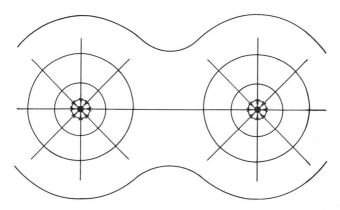

Fig. 10.3. Conceptual representation of flow lines and equipotentials surrounding two hypothetical roots. The flow lines converge toward the roots, where the gradient becomes steepest.

moisture regime, plants may be influenced more by the extreme values of the water potential they experience than by the average value (see Fig. 10.4). Furthermore, root distribution is not generally uniform or constant within the root zone. Nor does the water-extraction pattern necessarily correspond to the root-distribution pattern. Hence, the correlation of plant response with soil water often requires integrating each of the two over both space and time, so that the actual relationship becomes obscure.

Because of these complications, only a semiquantitative analysis is possible even at the present time.

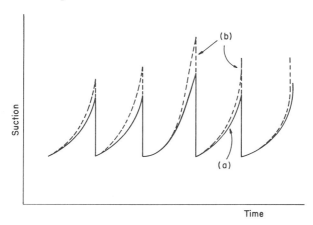

Fig. 10.4. The variation of soil water suction in the root zone during successive irrigation cycles: (a) the average suction (tensiometric measurement); (b) the suction of the soil in contact with the root.

C. Newer Concepts of Availability

In recent years, a fundamental change has taken place in the evaluation of soil–plant–water relationships. With the development of our theoretical understanding of the state and movement of water in the soil, plant, and atmosphere, and with the concurrent development of experimental techniques allowing more exact measurement of the interrelationships of potential, conductivity, water content, and flux both in the soil and in the plant, the way has been opened for a more basic approach to the problem. It has become increasingly clear that in a dynamic system such static concepts as the "soil-water constants" (i.e., "field capacity," "permanent wilting point," "critical moisture," "capillary water," "gravitational water," etc.) are often physically meaningless, as they are based on the supposition that processes in the field bring about static levels. In fact, as we are now very much aware, flow takes place almost incessantly, though in varying fluxes and directions, and static situations are exceedingly rare.

These developments have led to abandonment of the classical concept of "available water" in its original sense. Clearly, there is no fundamental qualitative difference between the water at one value of soil wetness and at another, nor is the amount and rate of water uptake by plants an exclusive function of the content or potential of soil water alone. The amount and rate of water uptake depend on the ability of the roots to absorb water from the soil with which they are in contact, as well as on the ability of the soil to supply and transmit water toward the roots at a rate sufficient to meet transpiration requirements. These, in turn, depend on *properties of the plant* (rooting density, rooting depth, and rate of root extension, as well as the physiological ability of the plant to increase its own water suction sufficiently to continue drawing water from the soil at the rate needed to avoid wilting), *properties of the soil* (hydraulic conductivity–diffusivity–matric suction–wetness relationships), and also to a considerable extent on the *micrometeorological conditions* (which dictate the rate at which the plant must transpire and hence the rate at which it must extract water from the soil in order to maintain its own hydration).

From a physical point of view, evapotranspiration can be viewed as a stream flowing from a *source* of limited capacity and of variable potential, namely the reservoir of soil moisture, to a *sink* of virtually unlimited capacity (though of variable evaporative potential)—the atmosphere. As long as the rate of root uptake of soil moisture balances the rate of canopy loss by transpiration, the stream continues at an undiminished rate. The moment the uptake rate falls below transpiration, the plant itself must begin to lose moisture. This imbalance cannot continue for long without resulting in loss of turgor and wilting of the plant.

Potential transpiration (Penman, 1949) is a measure of the rate water can be drawn from plant canopies by the atmosphere when the supply of soil water is not limiting.[2] An attempt has lately been made (Molz et al., 1968) to define a potential soil-moisture availability in terms of rate as the flux of water a soil can supply to a root site. At high values of soil wetness, actual transpiration is limited by and equal to the potential transpiration, which is largely dictated by external conditions. As soil wetness is depleted, transpiration can be limited by the potential soil-moisture availability. Much of the time, therefore, external meteorological conditions rather than soil conditions exercise the greatest influence over the transpiration rate.

D. Soil–Plant–Atmosphere System as a Physical Continuum

The current approach to the field water cycle is based on recognition that the field and all its components—soil, plant, and atmosphere taken together—form a physically unified and dynamic system (Gardner, 1960; Cowan, 1965) in which various flow processes occur interdependently like links in a chain. This unified system has been called the "SPAC" (for "soil–plant–atmosphere continuum") by J. R. Philip (1966). In this system, flow takes place from higher to lower potential energy, with the concept of "water potential" equally valid and applicable in the soil, plant, and atmosphere alike.

Unfortunately, workers in the fields of soil physics, plant physiology, and meteorology have long tended to express the same potential in different ways (such as "tension," "diffusion pressure deficit," "vapor pressure," etc.) and to measure these by the use of different units, so that the workers in these interrelated fields could not communicate with one another readily. The important principle to understand, therefore, is that the various terms used to characterize the state of water in different parts of the system are merely alternative expressions of the potential-energy level of the water, and that the difference in this level from one location to another is responsible for the tendency of the water to flow. This principle applies throughout the soil, the plant,[3] and the atmosphere continuously.

[2] When the plant canopy completely covers the soil surface, the potential transpiration rate is assumed to be equal to the *potential evapotranspiration* (or total evaporation) rate. As such, it is equivalent to the *external evaporativity* or the *evaporative demand*. The use of the word "potential" in this context, meaning "maximum possible rate," should not be confused with the potential-energy concept and the term "potential" denoting the energy state. Unfortunately, the same word is used in the literature for two such widely differing meanings.

[3] This statement is oversimplified, since plants tend to exclude salt and therefore the osmotic potential is added to the matric potential of soil water. Osmotic potential does not govern flow in most soils (which are not "salt sieving").

To characterize the SPAC physically, therefore, it is necessary to evaluate the energy potential of water, and its change with distance and time along the entire path of water movement. The flow rate is everywhere inversely proportional to a resistance. The flow path includes the water movement in the soil toward the roots, absorption into the roots, transport in the roots to the stems and through the xylem to the leaves, evaporation in the intercellular air spaces of the leaves, vapor diffusion through the stomatal cavities and openings to the quiescent air layer in contact with the leaf surface, and through it to the turbulent boundary layer, whence the vapor is finally transported to the external atmosphere.

The quantity of water transpired daily is large relative to the change in water content of the plant, so we can treat flow through the plant, for short periods, as a steady-state process (Gardner, 1960). If so, the potential differences in different parts of the system are proportional to the resistance to flow. This resistance is generally greater in the soil than in the plant, and greatest in the transition from the leaves to the atmosphere, where water changes its state from liquid to vapor and must move out by the relatively slow process of diffusion. The total potential difference between the soil and the atmosphere can amount to hundreds of bars, and in an arid climate can

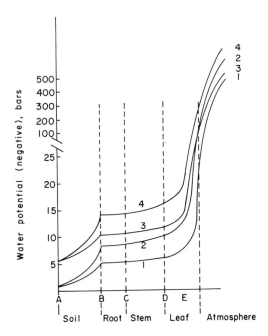

Fig. 10.5. The distribution of potentials in the soil–plant–atmosphere continuum under different conditions of soil moisture and atmospheric evaporativity.

even exceed 1000 bars. Of this total, the potential drop in the soil and plant is generally of the order of several bars to several tens of bars, so that the major portion of the overall potential difference in the SPAC occurs between the leaves and the atmosphere (Philip, 1966).

Figure 10.5 describes the distribution of water potentials in the SPAC. This figure is not drawn to scale and its purpose is only to illustrate general relationships. Curve 1 is for a low value of water suction in the soil (*AB*) and hence also at the root surfaces (*B*). In the mesophyl cells (*DE*), water suction (negative potential) is below the critical value at which the leaves may lose their turgidity, hence the plant is able to transport water from the soil to the atmosphere without wilting. E represents the substomatal cavity. In curve 2, soil water suction is equally low but the transpiration rate is higher and water suction in the mesophyl of the leaves approaches the critical wilting values (say, ~ 20 bars). Curve 3 is for the case in which soil water suction is relatively high but the transpiration rate is low. Curve 4, finally, indicates the extreme condition in which soil water suction and the transpiration rate are both high, the leaf water suction exceeds the critical value, and the plant wilts.

The resistance R_e to flow in each of the sections described[4] can be defined by the equation

$$R_e = -\frac{\Delta\phi}{q} \tag{10.1}$$

where $\Delta\phi$ is the potential drop and q the flux. Since this equation is mathematically identical to Ohm's law, it is possible to represent water flow in the SPAC by analogy with the current of electricity through a series of resistors. This analogy is illustrated in Fig. 10.6. The resistance of the soil to water flux, which can be defined as the ratio of the flow path length to the hydraulic conductivity, is shown as a variable resistance. This is because the hydraulic conductivity of the soil varies with changing soil wetness, and because the roots can increase their density within the existing rooting zone (thus decreasing the average flow path in the soil) as well as extend this zone by invading additional and moister soil layers. R_e depends, therefore, both on the soil moisture content and on the characteristics of the root system.

An alternative expression of Eq. (10.1) was given by Gardner and Ehlig (1963):

$$\phi_p = \phi_s - ER \tag{10.2}$$

---

[4] Strictly speaking, this representation of resistance may not apply to the transfer of water from the leaves to the atmosphere, as this stage of the process may be affected by an influx of external energy (e.g., radiation) rather than by the "internal" potential drop alone.

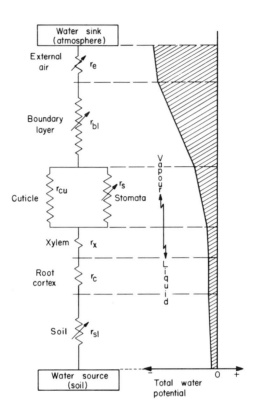

Fig. 10.6. An electrical analog representing resistances to flow in the SPAC. The resistances in the soil, leaf, and atmosphere may vary with variation of soil and meteorological conditions. Flow in the atmosphere is due to the vapor-pressure gradient and the transfer coefficient is a complex function of such variables as wind, turbulence etc. (after Rose, 1966).

where ϕ_p is water potential in the plant leaf, ϕ_s is water potential in the soil, E is transpiration rate, and R is resistance to flow in the combined soil–plant system. Since the potential and resistance values are in fact not constant but complex, these equations are more useful conceptually than quantitatively.

The resistance of the stomates is also variable, as they open, shut, or vary their aperture with changing leaf water potential. The atmospheric boundary layer may also vary its resistance, in accordance with the amount of mixing caused by turbulence. Resistance to water flow can vary within the plant also, but the magnitude of this variation is usually considered to be small.

As long as the plant does not wilt, and as long as the influx of radiation and heat to the canopy results in change of phase only, it is possible to assume steady-state flow through the plant. This means that the transpiration rate is equal to the plant transport and to the soil-water uptake rate:

$$q = -\frac{\Delta\phi_1}{R_1} = -\frac{\Delta\phi_2}{R_2} = -\frac{\Delta\phi_3}{R_3} \qquad (10.3)$$

where $\Delta\phi_1$ is the potential drop in the soil toward the roots, $\Delta\phi_2$ the potential drop in the plant to the leaves, and $\Delta\phi_3$ the potential drop between the leaves and the atmosphere. The magnitudes of these potential increments are of the order of: $\Delta\phi_1 \approx 10$ bars, $\Delta\phi_2 \approx 10$ bars, $\Delta\phi_3 \approx 500$ bars. It follows that the resistance R_{e3} between the leaves and the atmosphere might be 50 or more times greater than the resistances of the plant and the soil. When the stomates close, as at noontime during a hot day, the resistance R_{e3} may become even much greater, and result in a decreased rate of transpiration.

E. Flow of Water to Plant Roots

The root system of a plant can be quite extensive, with a total length of perhaps several miles. The total root area of an annual grass plant may be of the order of, say, 1000 m². If such a root system pervades a soil volume of about 100 liters, then, despite their considerable total length and surface, the roots can only come into direct contact with less than 1% of the particle surface area of, say, a medium-textured soil. It follows that water must move a considerable average distance in the soil before arriving at the nearest root surface. This distance can be of the order of millimeters or even centimeters, depending upon the density of roots and upon soil-water properties.

Thorough discussions and analyses of soil-water flow to plant roots were published by Gardner (1960, 1964, 1968). Much of our discussion here is based on his work.

Soil water suction increases as soil wetness decreases. Consequently, the plant water suction required to extract water from the soil must increase correspondingly. The soil will deliver water to the root as long as the water suction in the latter is maintained greater than in the former. However, as a root extracts water from the soil in contact with it, the suction in the contact zone may increase and tend to equal the root suction. In this case, water uptake might cease, unless additional water can move in from the farther reaches of the soil as a result of the soil-water-suction gradients which form toward the soil in direct contact with the root. In order for this additional water to become available to the plant, it must not only be at a suction lower than the root water suction, but must also move toward and into the root at a rate sufficient to compensate the plant for its own constant loss of water to the atmosphere by the process of transpiration. Clearly, therefore, the movement of water in the soil is a vital link in the chain of processes involved in the water economy of the plant. Furthermore, this movement must obey the general laws governing flow in an unsaturated soil.

On the assumptions that a typical root can be represented by an infinitely long, narrow cylinder of constant radius and absorbing characteristics (effectively a line sink), and that soil-water movement toward the root is radial, the appropriate form of the flow equation is

$$\frac{\partial \theta}{\partial t} = \frac{1}{r} \frac{\partial}{\partial r}\left(rD \frac{\partial \theta}{\partial r}\right) \tag{10.4}$$

where θ is the volumetric soil wetness, D diffusivity, t time, and r radial distance from the axis of the root. Assuming constant flux at the root surface, Gardner solved this equation subject to the following initial and boundary conditions:

$$\theta = \theta_0, \qquad \psi = \psi_0, \qquad t = 0$$

$$q = 2\pi a K \frac{\partial \psi}{\partial r} = 2\pi a D \frac{\partial \phi}{\partial r}, \qquad r = a, \qquad t > 0 \tag{10.5}$$

where a is the root radius, ψ matric suction, K conductivity, and q rate of water uptake per unit length of root.

With K and D assumed constant, the following solution is obtained:

$$\psi - \psi_0 = \Delta\psi = \frac{q}{4\pi K}\left(\ln \frac{4Dt}{r^2} - \gamma_c\right) \tag{10.6}$$

where γ_c is Euler's constant. From this equation, it is possible to calculate the gradient $\Delta\psi$ that will develop at any time between the soil at a distance from the root (i.e., the initial soil water suction ψ_0) and the suction ψ at the root-to-soil contact zone. Equation (10.6) shows that this gradient is proportional directly to the rate of water uptake and inversely to the hydraulic conductivity. The root water suction can therefore be expected to depend on these two factors as well as on the average soil water suction. Since the diffusivity D appears in the logarithmic term, Gardner concluded that variations in D can cause only slight changes in $\Delta\psi$, and hence the assumption of a constant D does not introduce a serious error. Also, the variation in K due to $\Delta\psi$ was considered to be no larger than the uncertainties in determination of K, so that the assumption of a constant K was valid for not too large t.

Equation (10.6) indicates that the distance at which any value of suction occurs increases as the square root of time. Thus, the distance from which water may eventually move to the root can be estimated from the time allowed. This analysis suggests that soil water can move toward the roots over distances of the order of several millimeters or even a few centimeters.

The dependence of suction on distance from the root during water uptake was calculated by Gardner for several soils and was found to be quite flat until the initial suction approached 15 bars. This is illustrated in Fig. 10.7.

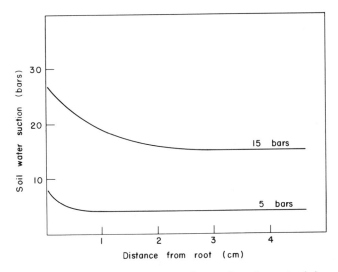

Fig. 10.7. Relation of soil water suction to distance from the root axis in a sandy soil with an uptake rate of 0.1 cm³ per cm of root per day. The two curves are for two different levels of soil water suction at a distance from the root. (After Gardner, 1960.)

The suction difference between soil and root ($\Delta\psi$) needed to maintain a steady flow rate depends, as stated, on the conductivity K and the flow rate q, which is the water-extraction rate required of the root by the transpiration process. When soil water suction is low and conductivity high, $\Delta\psi$ is small and the suction in the root will not differ markedly from the suction in the soil. When soil water suction increases, and soil conductivity decreases, the suction difference (or gradient) needed to maintain the same flow rate must increase correspondingly. As long as the transpiration rate required of the plant is not too high, and as long as the hydraulic conductivity of the soil is adequate and the density of the roots is sufficient, the plant can extract water from the soil at the rate needed to maintain normal activity. However, the moment the rate of extraction drops below the rate of transpiration (either because of a high evaporative demand by the atmosphere, and/or because of low soil conductivity, and/or because the root system is too sparse), the plant necessarily loses water, and, if it cannot adjust its root water suction or its root density so as to increase the rate of soil-water uptake, the plant may suffer from loss of turgor and be unable to grow normally. This situation will sooner or later cause the plant to wilt. It follows that, as atmospheric evaporativity increases, and as soil conductivity decreases, the average soil-water suction at which the plant wilts will tend to be lower.

Figure 10.8 shows the hypothetical relation of suction at the root to the average soil water suction for different rates of water uptake. The curves show that root water suction need not greatly exceed average soil water

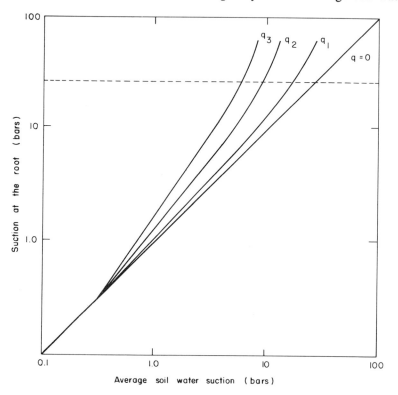

Fig. 10.8. The relation of suction at the root to the average soil water suction, for different uptake rates ($q_1 < q_2 < q_3$). The dashed line indicates an approximate value of critical suction at which the plant wilts. The soil water suction at which wilting occurs is seen to depend upon the uptake, and hence the transpiration, rate. (After Gardner, 1960.)

suction as long as the latter does not exceed a few bars. However, as average soil water suction increases beyond 10 bars or so, the suction which must develop at the root surface to maintain the flow rate constant may be 20 or 30 bars higher, and the difference must increase as the flux increases.

The theoretical dependence of water suction at the root upon the average soil moisture content for soils of different hydraulic properties is shown in Fig. 10.9. A study of this figure can help us to understand why, in a sandy soil, decreasing the water content has relatively little influence upon plant

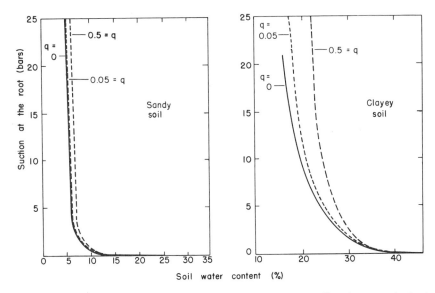

Fig. 10.9. Relation of water suction at the root to average soil moisture content, at various water uptake rates and for two different types of soil (after Gardner, 1960).

response until a critical point is reached, beyond which wilting takes place. The solid curves of this figure represent the suctions prevailing at the root surface when the uptake rate is zero. In other words, the solid curves are merely the familiar soil-moisture-retention curves for static equilibrium. In a sandy soil, most of the water is held at low suction values, and only the last 6–7 % of soil water is held in the range in which a steep rise in suction is needed for any further significant extraction of soil water to take place. On the other hand, in the clayey soil, suction increases more gradually with decreasing water content, and at any particular suction, the soil retains more moisture than a sandy soil. For this reason, the hydraulic conductivity of the clayey soil tends to be greater than that of the sandy soil for suctions exceeding, say, 1 bar. In a sandy soil, therefore, a change in the transpiration rate can hardly affect the wilting point significantly, while in a clayey soil, the wilting point can be strongly affected by the transpiration rate. This brings us back to the old controversy about "available water" and casts some light on it.

Assuming that wilting occurs at a certain limiting value of root water suction (e.g., 20 bars), it is possible to calculate the dependence of average soil wetness at which wilting occurs upon the transpiration (or uptake) rate. Figure 10.10 shows that the range of soil wetness at which plants might wilt is relatively narrow in a sandy soil and wider in a clayey soil (e.g., in Gardner's data, the wilting point of Chino clay increased from 16 to 23 %

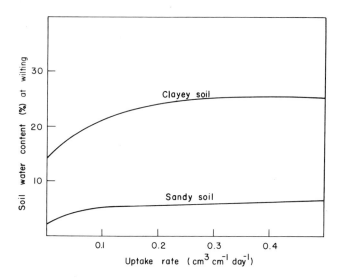

Fig. 10.10. Relation of soil wetness at wilting to the root uptake flux.

as the uptake rate increased from less than 0.1 to 0.5 ml/cm per day). Thus, both the suction and wetness at wilting are affected by the dynamics of water uptake.

F. Water Uptake by Root Systems

The discussion thus far has been based on the assumption that the roots are uniformly distributed in the rooting zone, and that average soil water suction is similarly uniform within the rooting zone. In actual fact, root systems in the field are seldom, if ever, uniform with depth. In general, moreover, the soil itself varies with depth.

As we have already stated, the rate of water uptake from a given volume of soil depends on rooting density (the effective length of roots per unit volume of soil), soil conductivity, and the difference between average soil water suction and root suction. If the initial soil water suction is uniform throughout all depths of the rooting zone, but the active roots are not uniformly distributed, the rate of water uptake should be highest where the density of roots is greatest. However, more rapid uptake will result in more rapid depletion of soil moisture, and the rate will not remain constant very long. If the distribution of roots with depth, the properties of the soil, and the initial water content are all known, it should be possible to predict the relative rate of water extraction from different depths in the soil. Such a calculation, carried out by Gardner (1964), is shown in Fig. 10.11.

Nonuniformity of water uptake from different soil depths has been found in the field, as shown in Fig. 10.12 (after Ogata *et al.*, 1960). This may be due to differences in the activity or conductance of different roots, as well as to differences in rooting density within the soil profile.

In a nonuniform root system, suction gradients can form which may induce water movement from one layer to another in the soil profile itself. In general, the magnitude of this movement is likely to be small relative to

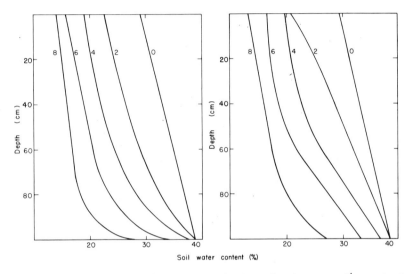

Fig. 10.11. Soil moisture profiles 0,2,4,6, and 8 days after the start of transpiration. At right: for a system with root density decreasing logarithmically with depth. At left: root density decreasing linearly with depth (after Gardner, 1964).

the water uptake rate by the plants, but in some cases it can be considerable. As a rough approximation, it is sometimes possible to divide the rooting system into two layers: an *upper layer*, in which root density is greatest and nearly uniform and in which water depletion is similarly uniform; and a *lower layer*, in which the roots are relatively sparse and in which the rate of water depletion is slow as long as the water content of the upper layer is fairly high. The water content of the lower layer is depleted by two sometimes simultaneous processes: uptake by the roots of that layer, and direct upward flow toward the more rapidly drying upper layer (caused by soil suction gradients).

A mature root system occupies a more-or-less constant soil volume of fixed depth so that uptake should depend mainly upon the size of this volume, its water content and hydraulic properties, and upon the density of the roots. On the other hand, in young plants, root extension and advance into deeper

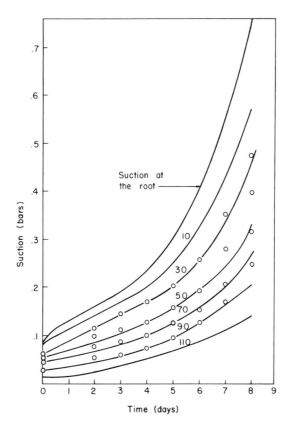

Fig. 10.12. The increase of average soil water suction with time in different depths in a field of alfalfa. The numbers by the curves represent different depths (cm) within the rooting zone. (After Ogata *et al.*, 1960.)

and moister layers can play an important part in supplying plant water requirements. This process has been studied by Kramer and Coile (1940) and by Wolf (1968).

For an idealized model in which: (a) the roots of a crop grow downward more or less in unison, forming an absorbing zone whose lower boundary approximates a moving plane at a constant diffusion pressure deficit; (b) the soil is initially uniform and of great depth; (c) the diffusivity of soil moisture is proportional to a positive power of the moisture content; and (d) no remoistening of the soil takes place; Wolf (1968) has shown that a steady state is established, typically within a few weeks, in which the maximum rate at which moisture can be acquired, per unit area of boundary, is proportional to the wetness of the soil and to the downward velocity of the root zone, and is independent of the diffusivity. With a growing root zone, the rate of water gain depends on the root water suction. Root growth also serves to bring a reserve of moisture into intimate contact with the roots, if the demand is not maximal.

In contrast, the flow to a stationary root zone declines with time, and if the exponent of the diffusivity function is large, increasing an already large root water suction does not materially change the flow. With growth, the total water made available in a few months is several times as large as without growth. This advantage increases with time, and is proportional to the growth velocity; it is less if the initial diffusivity is high, as in very moist soil.

One possible reason for the differences observed between the response of pot-grown vs. field-grown plants to the soil-water regime is the difference in root distribution with depth. In a pot, root density can be fairly uniform, while in the field, it generally varies with depth. Furthermore, the roots present in different layers may exhibit different water uptake and transmission properties. For instance, the roots of the deeper layers may offer greater resistance to water movement within the plant than the roots of the upper layers (e.g., Wind, 1955). The contribution of moist sublayers underlying the rooting zone can be decisive, especially where a high water table is present.

Rose and Stern (1967) presented an analysis of the time rate of water withdrawal from different soil depth zones in relation to soil wetness and hydraulic properties, and to the rate of plant-root uptake.

The water-conservation equation for a given depth of soil (assuming flow to be vertical only) for a given period of time (t_1 to t_2) can be written

$$\int_{t_1}^{t_2} (i - v_z - q_e) \, dt - \int_0^z \int_{t_1}^{t_2} \left(\frac{\partial \theta}{\partial t} \right) dz \, dt = \int_0^z \int_{t_1}^{t_2} r_z \, dz \, dt. \qquad (10.7)$$

where i is rate of water supply (precipitation or irrigation), q_e evaporation rate from the soil surface, v_z the vertical flux of water at depth z, θ is volumetric soil wetness, and r_z the rate of decrease of soil wetness due to water uptake by roots.

The average rate of uptake by roots at the depth z is

$$\bar{r}_z = \int_{t_1}^{t_2} r_z \, dt / (t_2 - t_1) \qquad (10.8)$$

The pattern of soil-water extraction by a root system can be determined by repeated calculations based on the above equations for successive small intervals of time and depth. The total (cumulative) water uptake by the roots R_z is given by

$$R_z = \int_0^z r_z \, dz \qquad (10.9)$$

The relationships were used to describe the pattern of soil-water extraction by a cotton crop in the field. The results indicated that nearly all water extraction by the crop took place from the top 30 cm during the early stages of growth and from the top 100 cm during later stages.

A similar and detailed field study of water uptake by a root system was carried out by van Bavel *et al.* (1968a,b), using the *instantaneous profile method* of obtaining the hydraulic properties of a complete profile *in situ* (Watson, 1966). (This method requires frequent, independent, and simultaneous measurements of hydraulic head and soil wetness throughout the profile, coupled with measurements of evaporation.) The calculated root extraction rates agreed reasonably well with separate measurements of transpiration obtained with lysimeters. Soil-water movement within the root zone of a sorghum crop indicated initially a net downward outflow from the root zone but this movement later reversed itself to indicate a net upward inflow from the wet soil to the root zone above.

G. Interaction of Soil Wetness, Suction, and Transpiration Rate

As we have already pointed out, when plant canopies cover the soil surface and soil water suction is relatively low, the transpiration rate generally approaches the potential evapotranspiration rate determined by meteorological factors.

The plant, as well as the soil, is a hydraulic conductor, in which the rate of flow is proportional to the product of a driving force (a potential or suction gradient) and a transmission coefficient (conductance). If the conductance of the plant remains constant despite changes in water potential, then the suction difference between roots and leaves depends only upon the transpiration rate. However, plant conductance apparently does not remain constant. In particular, a progressive reduction of leaf-water potential eventually results in closure of the stomates and increase of resistance to water transport to the atmosphere.

The actual relation of leaf-water potential to stomatal aperture and its effects upon the processes of transpiration, synthesis, respiration, and growth—all these belong in the realm of plant physiology and are outside the scope of our necessarily limited text.[5] In any case, it should be clear that continued increase of soil moisture suction must sooner or later result in a decrease of the transpiration rate, and this decrease sometimes occurs fairly abruptly just before the plant wilts. A clayey soil can maintain a high rate of transpiration longer than a sandy soil, since both its water content and

[5] The reader may refer to Slatyer (1967) or to Kozlowski (1968).

hydraulic conductivity are generally higher in the unsaturated state (Gardner and Ehlig, 1962).

Gingrich and Russell (1967) compared the growth of corn seedlings grown in osmotic solutions to seedlings grown in soil of known matric suction. They found that increasing soil water suction had a greater effect in retarding transpiration and growth than increasing osmotic suction in substrate solutions. Their findings are indicative of the possible effect of the hydraulic transmission characteristics (i.e., the hydraulic conductance or its reciprocal, the resistance) which can become limiting in the case of a desorbing soil but not in the case of a solution.

Denmead and Shaw (1962) have found experimental confirmation of the effect of dynamic conditions on water uptake and transpiration. They measured the transpiration rates of corn plants grown in containers and placed in the field under different conditions of irrigation and atmospheric evaporativity. Under an evaporativity of 3–4 mm/day, the actual transpiration rate began to fall below the potential rate at an average soil water suction of about 2 bars. Under more extreme meteorological conditions, with an evaporativity of 6–7 mm/day, this drop occurred already at a soil water suction value of 0.3 bar. On the other hand, when the potential evaporativity was very low (1.4 mm/day), no drop in transpiration rate was noticed until average soil water suction exceeded 12 bars. The volumetric water contents at which the transpiration rates fell varied between 23%, under the lowest evaporativity, and 34%, under the highest evaporativity measured. This is illustrated in Figs. 10.13 and 10.14.

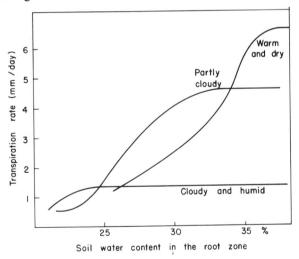

Fig. 10.13. Relation of actual transpiration rate to soil water content, under different meteorological conditions. (After Denmead and Shaw, 1962.)

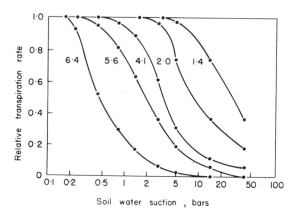

Fig. 10.14. The relation of relative transpiration rate to average soil water suction, under different meteorological conditions. The numbers represent different rates of potential evapotranspiration. (After Denmead and Shaw, 1962.)

H. Transpiration in Relation to Production

The aim of soil and water management is generally to maximize production in relation to investment (e.g., in land, water, labor, machinery, energy, fertilizers, etc.). In arid regions, where water generally constitutes the limiting factor, the aim is generally to induce the greatest amount of growth per unit quantity of available water. A shortage of water, particularly if it occurs for long periods or during critical stages of growth, can obviously curtail crop yields. On the other hand, an excessive supply of water can be not only wasteful but harmful as well (as it might hinder growth by limiting aeration, leaching nutrients, increasing salinity, etc.).

To characterize the relation of total seasonal water use by a crop to its dry matter production, early investigators (e.g., Briggs and Shantz, 1913, 1914) introduced the term *transpiration ratio*. This index was found to depend primarily on the climate, and to range up to several hundred (or even one thousand or more) in arid regions. Later, the inverse concept of *water use efficiency* came into use, defined (Viets, 1962) as the ratio of the dry matter production to the amount of water transpired by a crop in the field.

Transpiration can be limited either by the supply of water to be evaporated (by precipitation or irrigation), or by the supply of energy needed for vaporization, the latter determined mainly by such climatic factors as radiation, temperature, and humidity. When the supply of water is plentiful, it is the climate, and particularly the net radiation, which determine water use (Penman, 1956). When water is limiting, the assimilation rate, plant growth, and consequently crop yield are all related quantitatively to the water supply, but this relationship may not be a simple one. An excellent comprehensive review and analysis of this relationship in temperate and arid climates was published some years ago by de Wit (1958). He found that in climates with a large percentage of bright sunshine (i.e., arid regions) a relation

$$Y_d = mE_a/E_0 \qquad (10.10)$$

exists between total dry matter yield (Y_d) and the ratio of actual transpiration (E_a) and free water evaporation (E_0). In climates with a small percentage of bright sunshine (i.e., temperate regions) the relation

$$Y_d = nE_a \qquad (10.11)$$

was found (that is to say, dry-matter production is proportional to transpiration and hence water use efficiency is constant). These relations were found for both container-grown and field-grown plants, and the values of constants m and n were reported to be characteristic of the crops. These relations obviously cannot hold as potential evapotranspiration (to be discussed in our next chapter) is approached and water ceases to be the limiting factor in plant growth. Beyond the point where transpiration reaches its climatic limit, the promise of increasing production lies in obviating the other possible limiting factors, such as light intensity, carbon dioxide concentration in the air, nutrients in the soil, etc.

I. Summary

The state and movement of water in the soil, plant, and atmosphere are affected by a complex set of interactions and of processes which occur simultaneously at different rates. For many years, research in this area was hampered by a failure to appreciate the overall physical unity and dynamic nature of the system. The literature abounded with a mass of seemingly conflicting experimental data, variously interpreted. The results obtained from empirical work alone may depend on uncontrollable, and often unrecognized, variables. Only a comprehensive physical understanding of the soil–plant–atmosphere system as a whole can help us to avoid drawing wrong conclusions and unwarranted generalizations from a specific set of experimental findings.

The physical analysis of the processes constituting the field water cycle is still in its beginning stages, and is as yet based upon simplifying assumptions which do not always represent the complex conditions which in fact prevail at different locations. However, it already appears possible to formulate physically, in quantitative terms, processes which until recently could be described only qualitatively.

Water movement from the soil, through the plant, and to the atmosphere occurs along a path of continuously decreasing potential energy. This path includes a number of distinct segments, each of which can be described in terms of a flow equation. The first link in this chain is the flow of water in the unsaturated soil surrounding the root, and as such it can affect the overall flow process. The ability of a plant to obtain a sufficient water supply depends not only on the content and potential of soil water, but also upon the required flow rate and transmitting properties of the soil.

From the dynamic point of view, the popular and oft-abused concepts of "field capacity" and "wilting point" must be redefined, if they are to be used at all, not as static levels of soil moisture, but in terms of process rates. Soil wetness can be considered to be below "field capacity" when the rate of internal drainage from the root zone is small enough to be disregarded, for a time, under a specific set of circumstances. The "wilting point" can be defined as the soil wetness below which soil-water extraction by a given plant is insufficient to balance the transpiration rate demanded of it by the atmosphere, in a specific climatic environment.

Plant response to the soil-moisture regime depends also on a large number of environmental variables not included in this discussion. Such *soil factors* as aeration, nutrient availability, and mechanical properties, such *plant factors* as age and various genetic traits, and such *meteorological factors* as light intensity, day length and atmospheric composition can influence not only the magnitude of plant response to soil water, but also the direction of this response.

11 *Water Balance and Energy Balance in the Field*

A. General

The various soil-water flow processes which we have attempted to describe separately (e.g., infiltration, redistribution, drainage, evaporation, and water uptake by plants) are in fact strongly interdependent, as they occur sequentially or simultaneously. To evaluate the field water cycle as a whole, and the relative magnitudes of the various processes comprising it over a period of time, it is necessary to consider the *field water balance*. Just as a businessman regularly summarizes the financial balance of his enterprise, including an itemized listing of all income sources, expenditures, inventory changes, and net worth, so the agricultural or environmental physicist can attempt to account for all the water entering, leaving, and remaining in a specified volume of soil during a specified length of time.

The water balance is merely a detailed statement of the *law of conservation of matter*, which states simply that matter can neither be created nor destroyed but can only change from one state or location to another. Thus, the water content of a given soil volume cannot increase without addition from the outside (as by infiltration or capillary rise), nor can it diminish unless transported to the atmosphere by evapotranspiration or to deeper zones by drainage.[1]

The field water balance is intimately connected with the *energy balance*, since it involves processes that require energy. The energy balance is an expression

---

[1] We are disregarding the possibility of lateral flows.

of the classical *law of conservation of energy*, which states that, in a given system, energy can be absorbed from, or released to, the outside, and that along the way it can change form, but it cannot be created or destroyed.

The content of water in the soil affects the way the energy flux reaching the field is partitioned and utilized. Likewise, the energy flux affects the state and movement of water. The water balance and energy balance interact, since they are both aspects of the same processes within the same environment. A physical description of the soil–plant–atmosphere system, therefore, must be based on an understanding of both balances together. In particular, the evaporation process, which is often the principal consumer of both water and energy in the field, depends, in a combined way, on the simultaneous supply of water and energy.

B. Water Balance of the Root Zone

In its simplest form, the water balance merely states that, in a given volume of soil, the difference between the amount of water added W_{in} and the amount of water withdrawn W_{out} during a certain period is equal to the change in water content ΔW during the same period:

$$W_{in} - W_{out} = \Delta W \tag{11.1}$$

When gains exceed losses, the water-content change is positive; and conversely, when losses exceed gains, ΔW is negative.

In order to itemize the equation, consider the fate of rain or irrigation water reaching the soil surface during a given period of time. Some of this water infiltrates into the soil and some may accumulate temporarily in puddles over the surface or trickle downslope as surface runoff. Of the water entering the soil, some may evaporate directly through the soil surface, some may be extracted by plants and transpired to the atmosphere, some may drain out of the soil depth considered, and the remainder (at a given time stage) is necessarily stored within the soil. Additional water can also reach the field, either by runoff from an adjacent field, or by capillary rise from a water table or from wet layers present at some depth. The pertinent soil depth for which the water balance is computed is determined arbitrarily. Thus, in principle, a water balance can be computed for a small sample of soil or for an entire watershed. From an agricultural or plant ecological point of view, it is generally most appropriate to consider the rooting depth (or root zone) of the plants growing in the field.

We shall now proceed to itemize the income and outgo components of the field water balance.

The amount of water added W_{in} may consist of precipitation M or irrigation I_r, or both. These are generally expressed in units of volume per unit area—i.e., in units of head, or depth, of water:

$$W_{in} = M + I_r \qquad (11.2)$$

At the same time, water may be withdrawn from the soil (W_{out}) by the processes of runoff, drainage, and evapotranspiration. Thus:

$$W_{out} = N + F + (E + T) \qquad (11.3)$$

where N is runoff (which generally involves a loss of water by the field, but may constitute a gain if runoff is entering "our" field from an adjacent one); F is "deep percolation" or drainage out of the root zone (this, too, may constitute either a positive or a negative increment of water, depending on whether flow is upward or downward); E is direct evaporation from the soil surface; T is transpiration from vegetation; and therefore $(E + T)$ is the evapotranspiration. In Eq. (11.3), we have disregarded interception of water by plant canopies and surface storage in puddles, since these phenomena are generally very temporary and the water thus detained generally ends up as delayed infiltration or as evaporation.

The total water balance is, accordingly,

$$\Delta W = M + I_r - N - F - (E + T) \qquad (11.4)$$

This is an integral form of the water balance, with the various items totalled over a certain period of time. The field water balance can also be expressed in differential form, referring to the time-rates of the various simultaneous processes (i.e., the simultaneous fluxes across a plane). The sums of the separate terms must obey the mass-conservation law in both the integral and differential forms.

The various items entering into the water balance of a hypothetical root zone are illustrated in Fig. 11.1. In this representation, only vertical flows

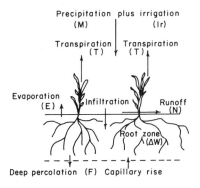

Fig. 11.1. The water balance of a root zone (schematic).

are considered within the soil. In a larger sense, any soil layer of interest forms a part of an overall hydrologic cycle, illustrated in Fig. 11.2, in which the flows are multidirectional.

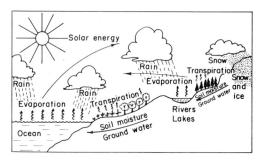

Fig. 11.2. The hydrologic cycle (schematic). (After Bertrand, 1967.)

C. Evaluation of the Water Balance

Simple and readily understandable though the field water balance may seem in principle, it is still rather difficult to measure in practice. A single equation can be solved if it has only one unknown. Often the largest component of the field water balance, and the one most difficult to measure directly, is the evapotranspiration $(E + T)$, also designated E_t. To obtain E_t from the water balance (Deacon *et al.*, 1958) we must have accurate measurements of all other terms of the equation. It is relatively easy to measure the amount of water added to the field by rain and irrigation. $(M + I_r)$, though it is necessary to consider possible nonuniformities in areal distribution. The amount of runoff generally is (or at least should be) small in agricultural fields, and particularly in irrigated fields, so that it can sometimes be regarded as neglible in comparison with the major components of the water balance.

For a long period, e.g., an entire season, the change in water content of the root zone is likely to be small in relation to the total water balance. In this case, the sum of rain and irrigation is approximately equal to the sum of evapotranspiration E_t and deep percolation F. For shorter periods, the change in soil water content ΔW can be relatively large and must be measured. This measurement can be made by sampling periodically, or by use of specialized instruments.[2]

---

[2] Of the various methods for measuring the content of water in soil, the neutron meter is the most satisfactory at present since it measures wetness on the volume or depth fraction basis directly and since it samples a large volume and minimizes sampling errors (with repeated measurements made at the same site and depth (Holmes *et al.*, 1967).

During dry spells, without rain or irrigation, $W_{in} = 0$, so that the sum of F and E_t now equals the reduction in root-zone water content:

$$-\Delta W = F + E_t \tag{11.5}$$

Common practice in irrigation is to measure the total water content of the root zone just prior to an irrigation, and to supply the amount of water necessary to replenish the deficit to some maximal water content, which is taken to be the "field capacity." Some ecologists and irrigationists have tended to assume that the deficit which develops between rains or irrigations is due to evapotranspiration only, thus disregarding the amount of water which may flow through the bottom of the root zone, either downward or upward. This flow is not always negligible, and can often constitute a tenth or more of the total water balance (Robins et al., 1954; Nixon and Lawless, 1960; Rose and Stern, 1967).

It should be obvious that measurement of root-zone or subsoil water content by itself cannot tell us the rate and direction of soil-water movement (van Bavel et al., 1968a,b). Even if the water content at a given depth remains constant, we cannot conclude that the water there is immobile, since it might be moving steadily through that depth. Tensiometric measurements can, however, indicate the directions and magnitudes of the hydraulic gradients through the profile (Richards, 1967) and allow us to compute the fluxes from knowledge of the hydraulic conductivity versus suction or wetness for the particular soil. More direct measurements of the deep percolation component of the field water balance will become possible with the development of water flux meters (Cary, 1968). Such devices are still in the preliminary development stage, however.

In irrigated agriculture, the ratio of E_t to the sum $E_t + F$ is sometimes used as the index of *irrigation efficiency*. This ratio can approach 100% when deep percolation is negligible ($F \approx 0$). Such a high efficiency is not always desirable, however, since it might result in progressive salinization of the soil, especially in arid regions. Plants normally extract water while leaving behind much of the soluble salts originally present in the soil solution. Excess irrigation is therefore needed to provide deep percolation and leaching, and thus to prevent accumulation of excess salts in the root zone. On the other hand, too low an irrigation efficiency is equally undesirable, since it wastes water and might also leach nutrients out of the root zone, impede aeration, and raise the water table. It follows that movement below the root zone can be an important factor in the field water budget, not to be disregarded, but to be measured and controlled.

The most direct method for measurement of the field water balance is by use of *lysimeters* (van Bavel and Myers, 1962; Pruitt and Angus, 1960; King et al., 1956; Pelton, 1961; McIlroy and Angus, 1963; Forsgate et al.,

1965; Rose et al., 1966; Harrold, 1966; Black et al., 1968; Hillel et al., 1969). These are generally large containers of soil, set in the field to represent the prevailing soil and climatic conditions, and allowing more accurate measurement of physical processes than can be carried out in the open field. From the standpoint of the field water balance, the most efficient lysimeters are those equipped with a weighing device and a drainage system, which together allow continuous measurement of both evapotranspiration and percolation. Lysimeters may not provide a reliable measurement of the field water balance, however, when the soil or above-ground conditions of the lysimeter differ markedly from the actual field environment.

D. Radiation Exchange in the Field

By *radiation* we refer herein to the emission of energy in the form of electromagnetic waves from all bodies above 0°K. *Solar* (sun) *radiation* received on the earth's surface is the major component of its energy balance. Green plants are able to convert a part of the solar radiation into chemical energy. They do this in the process of photosynthesis, upon which all life on earth ultimately depends. For these reasons, it is appropriate to introduce a discussion of the energy balance with an account of the radiation balance, which constitutes so important a part of it.

Solar radiation reaches the outer surface of the atmosphere at a nearly constant flux of about 2 cal min^{-1} cm^{-2} perpendicular to the incident radiation.[3] Nearly all of this radiation is of the wavelength range of 0.3–3 microns (3000–30,000 Å), and about half of this radiation consists of visible light (i.e., 0.4–0.7 micron in wavelength). The solar radiation corresponds approximately to the emission spectrum of a "black body"[4] at a temperature of 6000°K. The earth, too, emits radiation, but since its surface temperature is about 300°K, this *terrestrial radiation* is of much lower intensity and greater wavelength than solar radiation[5] (i.e., in the wavelength range of 3–50

---

[3] 1 cal/cm^2 = 1 langley (Ly). 58 Ly ≈ 1 mm evaporation equivalent (latent heat = 580 cal/gm).

[4] A "black body" is one which absorbs all radiation reaching it without reflection, and emits at maximal efficiency. According to the *Stefan–Boltzmann law*, the total energy emitted by a body J_t, integrated over all wavelengths, is proportional to the fourth power of the absolute temperature T. This law is usually formulated as $J_t = \varepsilon \sigma T^4$ (where σ is a constant, and ε the emissivity coefficient). For a perfect black body, $\varepsilon = 1$.

[5] According to *Wien's law*, the wavelength of maximal radiation intensity is inversely proportional to the absolute temperature: $\lambda_m T = 2900$ (where λ_m is the wavelength in microns and T is the temperature on the Kelvin scale). *Planck's law* gives the intensity-distribution of energy emitted by a black body as a function of wavelength and temperature: $E_\lambda = C_1/\lambda^5 [\exp(C_2/\lambda T) - 1]$ (where E_λ is the energy flux emitted in a particular wavelength range, and C_1, C_2 are constants).

microns). Between these two radiation spectra, the sun's and the earth's, there is very little overlap, and it is customary to refer to the first as "short-wave" and to second as "long-wave" radiation (Sellers, 1965).

In passage through the atmosphere, solar radiation changes both its flux and spectral composition. About one-third of it, on the average, is reflected back to space (this reflection can become as high as 80% when the sky is completely overcast with clouds). In addition, the atmosphere absorbs and scatters a part of the radiation, so that only about half of the original flux density of solar radiation finally reaches the ground.[6] A part of the reflected and scattered radiation also reaches the ground and is called *sky radiation*. The total of direct solar and sky radiations is termed *global radiation*.

Albedo is the reflectivity coefficient of the surface toward short-wave radiation. This coefficient varies according to the color, roughness, and inclination of the surface, and is of the order of 5–10% for water, 10–30% for a vegetated area, 15–40% for a bare soil, and up to 90% for snow.

In addition to these incoming and reflected short-wave radiation fluxes, there is also a long-wave radiation (heat) exchange. The earth's surface emits radiation, and at the same time the atmosphere absorbs and emits long-wave radiation, part of which reaches the surface. The difference between the outgoing and incoming fluxes is called the *net long-wave radiation*. During the day, the net long-wave radiation may be a small fraction of the total radiation balance, but during the night, in the absence of direct solar radiation, the heat exchange between the land surface and the atmosphere dominates the radiation balance.

The overall difference between total incoming and total outgoing radiation (including both the short-wave and long-wave components) is termed *net radiation*, and it expresses the rate of radiant energy absorption by the field.

$$J_n = J_s\!\downarrow - J_s\!\uparrow + J_l\!\downarrow - J_l\!\uparrow \tag{11.6}$$

where J_n is the net radiation; $J_s\!\downarrow$ the incoming flux of short-wave radiation from sun and sky; $J_s\!\uparrow$ the short-wave radiation reflected by the surface; $J_l\!\downarrow$ the long-wave radiation from the sky; and $J_l\!\uparrow$ the long-wave radiation reflected and emitted by the surface.

At night, the short-wave fluxes are negligible, and since the long-wave radiation emitted by the surface generally exceeds that received from the sky, the nighttime net radiation flux is negative.

---

[6] In arid regions, where the cloud cover is sparse, the actual radiation received at the soil surface can exceed 70% of the "external" radiation. In humid regions, this ratio can be 40% or lower.

The reflected short-wave radiation is equal to the product of the incoming short-wave flux by the reflectivity coefficient (the albedo, α_r):

$$J_s\uparrow = \alpha_r J_s\downarrow$$

Therefore,

$$J_n = J_s\downarrow (1 - \alpha_r) - J_l \tag{11.7}$$

where J_l is the net flux of long-wave radiation, which is given a negative sign (since the surface of the earth is usually warmer than the atmosphere, there is generally a net loss of thermal radiation from the surface). As a rough average, J_n is generally of the order of 55–70% of $J_s\downarrow$ (Tanner and Lemon, 1962).

E. Total Energy Balance

Having balanced the gains and losses of radiation at the surface to obtain the net radiation, we can now proceed with an accounting of the transformations of this energy.

Part of the net radiation received by the field is transformed into heat, which warms the soil, plants, and atmosphere. Another part is taken up by the plants in their metabolic processes (e.g., photosynthesis). Finally, a major part is absorbed as latent heat in the twin processes of evaporation and transpiration. Thus,

$$J_n = LE + A + S + M \tag{11.8}$$

where LE is the rate of energy utilization in evapotranspiration (a product of the rate of water evaporation E and the latent heat of vaporization L), A is the energy flux that goes into heating the air (called *sensible heat*), S is the rate at which heat is stored in the soil, water, and vegetation, and M represents other miscellaneous energy terms such as photosynthesis and respiration.

The energy balance is illustrated in Fig. 11.3.

Where the vegetation is short (e.g., grass or field crops), the storage of heat in the vegetation is negligible compared with storage in the soil (Tanner, 1960), except, perhaps, in the case of the voluminous vegetation of a forest. The heat stored in the soil under short and sparse vegetation may be a fairly large portion of the net radiation at any one time during the day, but the net storage over a 24-hr period is usually small (since the nighttime loss of soil heat negates the daytime gain). For this reason, mean soil temperature generally does not change appreciably from day to day. The soil-storage term has variously been reported to be of the order of 5–15% of J_n (Decker 1959; Tanner and Pelton, 1960). This obviously depends on season. In spring and summer, this term is positive, but it becomes negative in autumn.

In the past, the miscellaneous energy terms [M in Eq. (11.8)] were believed

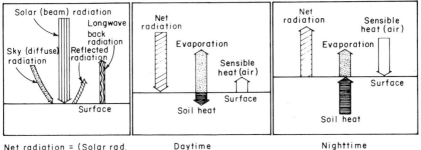

Net radiation = (Solar rad.   Daytime   Nighttime
+ sky rad.) - (Reflected rad.
+ back rad.)

Fig. 11.3. Schematic representation of the radiation balance and of the daytime and night-time energy balance. It is to be remembered that the daytime net radiation during the growing season is much greater than at night. (After Tanner, 1968.)

to be a negligible portion of the energy balance. Measurements of carbon dioxide exchange over active crops in the natural environment, however, have revealed that photosynthesis may account for as much as 5–10% of the daily net radiation where there is a large mass of active vegetation (particularly under low-light conditions, as when the sky is overcast). In general, though, M is less than 5% of J_n (Lemon, 1960).

Overall, the amount of energy stored in soil and vegetation and that fixed photochemically account for a rather small portion of the total daily net radiation, with the major portion going into latent and sensible heat. The proportionate allocation between these terms depends on the availability of water for evaporation, but in most agriculturally productive fields the latent heat predominates over the sensible heat term.

F. Transport of Heat and Vapor to the Atmosphere

The transport of sensible heat and water vapor (which carries latent heat) from the field to the atmosphere is affected by the turbulent movement of the air in the atmospheric boundary layer.[7] The sensible heat flux A is proportional to the product of the temperature gradient dT/dz by the turbulent transfer coefficient for heat, k_a (cm^2 sec^{-1}):

$$A = -c_p \rho_a k_a \frac{dT}{dz} \tag{11.9}$$

[7] "A laminar boundary layer", generally less than 1 mm thick, is recognized in immediate contact with the surface of an evaporating body. Through this layer, transport occurs by diffusion. Beyond this, turbulent transport becomes predominant in the "turbulent boundary layer".

where c_p is the specific heat capacity of air at constant pressure (cal gm^{-1} °C^{-1}), ρ_a the density of air, T temperature (°C), and z height (cm).

The rate of latent heat transfer by water vapor from the field to the atmosphere, LE, is similarly proportional to the product of the vapor pressure gradient by the appropriate turbulent transfer coefficient for vapor.

If we assume that the transfer coefficients for heat and water vapor are equal, then the ratio of the sensible heat transport to the latent heat transport becomes

$$\beta_0 = \frac{A}{LE} \approx \xi \frac{\Delta T}{\Delta e} \qquad (11.10)$$

where $\Delta T/\Delta e$ is the ratio of the temperature gradient to the vapor pressure gradient in the atmosphere above the field, and ξ is the psychrometric constant ≈ 0.66 mb/°C.

The ratio β_0 is called the *Bowen ratio*, and it depends mainly on the temperature and moisture regime of the field. When the field is wet, the relative humidity gradients between its surface and the atmosphere tend to be large, whereas the temperature gradients tend to be small. Thus, β_0 is rather small when the energy is consumed mainly in evaporation. When the field is dry, on the other hand, the relative humidity gradients toward the atmosphere are generally small, and the temperature gradients tend to be steep, so that the Bowen ratio becomes large. In a recently irrigated field, β_0 may be smaller than 0.2, while in a dry field in which the plants are under a water stress (with stomatal resistance coming into play), the surface may warm up and a much greater share of the incoming energy will be lost to the atmosphere directly as sensible heat. Under extremely arid conditions, in fact, LE may tend to zero and β_0 to infinity. With advection (section G), sensible heat can be transferred from the air to the field, and the Bowen ratio can become negative.

Whether or not water-vapor transport to the atmosphere from a vegetated field becomes restricted obviously depends not only upon the soil water content *per se*, but on a complex interplay of factors in which the characteristics of the plant cover (i.e., density of the canopy, root distribution, and physiological responses to water stress) play an important role.

The assumption that the transfer coefficients for heat and vapor are equal (or at least proportional) is known as the principle of *similarity* (Tanner, 1968). Transfer through the turbulent atmospheric boundary layer takes place primarily by means of *eddies*, which are ephemeral, swirling microcurrents of air, whipped up by the wind. Eddies of varying size, duration, and velocity fluctuate up and down at varying frequency, carrying both heat and vapor. While the instantaneous gradients and vertical fluxes of heat and vapor will generally fluctuate, when a sufficiently long averaging period is allowed (say, 15–60 min), the fluxes exhibit a stable statistical relationship over a uniform field.[8] This is not the case at a low level over an inhomogeneous surface (e.g., spotty vegetation and partially exposed soil). Under such conditions, cool, moist packets of air may rise from the vegetated spots while warm, dry air may rise from the dry soil surface, with the latter rising more rapidly owing to buoyancy.[9]

Using the Bowen ratio, the latent and sensible heat fluxes can be written (recalling that $J_n = S + A + LE$, and that $\beta_0 = A/LE$):

$$LE = (J_n - S)/(1 + \beta_0) \qquad (11.12)$$

$$A = \beta_0(J_n - S)/(1 + \beta_0) \qquad (11.13)$$

LE can thus be obtained from micrometeorological measurements in the field (i.e., J_n, S, and β_0) without necessitating measurements of soil-water fluxes or plant activity.

The diurnal variation of the components of the energy balance is illustrated in Fig. 11.4. The diurnal as well as the annual patterns of the components of the energy balance differ for different conditions of soil, vegetation, and climate (Sellers, 1965).

G. Advection

The equations given for the energy balance are applicable most readily to extensive and uniform areas in which the fluxes of energy are essentially vertical. Small fields can be influenced by their surroundings, and winds can often result in net horizontal transport of heat into or out of any particular field if it is not in the midst of a large and homogeneous region. This phenomenon, called *advection*, can be especially important in arid regions, where small, irrigated fields are often surrounded by an expanse of dry land, and the warm incoming air can transfer sensible heat down to the crop (Graham and King, 1961; Halstead and Covey, 1957).

The extraction of sensible heat from a warm mass of air flowing *over* the top of a field, and the conversion of this heat to latent heat of evaporation, is called the *oasis effect*. The passage of warm air *through* the vegetation has been called the *clothesline effect* (Tanner, 1957). A common sight in arid regions is the poor growth of the plants near the windward edge of a field,

---

[8] It is reasonable to assume that momentum and carbon dioxide, as well as vapor and heat, are carried by essentially the same eddies.

[9] An index of the relative importance of buoyancy (thermal) vs. frictional forces in producing turbulence is the Richardson number, $R_n = \rho(dT/dz)/(T(du/dz)^2)$, where dT/dz is the temperature gradient, du/dz is the wind velocity gradient, T is temperature, and ρ the mass of the gaseous component per volume of air. R_n, and hence the buoyancy effect, will be large only when the wind gradient is very small.

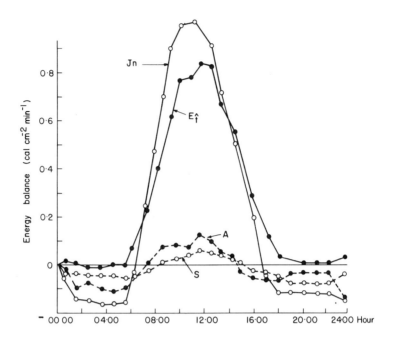

Fig. 11.4. The diurnal variation of net radiation J_n, and of energy utilization by evapotranspiration E_t, sensible heating of the atmosphere A, and heating of the soil S. Alfalfa-brome hay on Plainfield sand, September 4, 1957. (After Tanner, 1960).

where penetration of warm, dry wind contributes energy for evapotranspiration. Where advective heat inflow is large, evapotranspiration from rough and "open" vegetation (e.g., widely spaced tow crops or trees) can greatly exceed that from smooth and close vegetation (e.g., mowed grass).

The effects of advection are likely to be small in very large and uniform fields, but very considerable in small plots which differ from their surroundings. With advection, latent heat requirements can be larger than net radiation. Hence, values of evapotranspiration obtained from small experimental plots are not typically representative of large fields, unless these plots are guarded for some distance by an expanse, or *fetch*, of vegetation of similar roughness characteristics and subject to similar moisture conditions.

H. Potential Evapotranspiration (The Penman Equation)

The concept of *potential evapotranspiration* is an attempt to characterize the micrometeorological environment of a field in terms of an evaporative power, or demand; i.e., in terms of the maximal evaporation rate which the atmosphere is capable of exacting from a field of given surface properties. More specifically, Penman (1956) defined it as, "The amount of water transpired in unit time by a short green crop, completely shading the ground, of uniform height and never short of water." As such, it is a useful standard of reference for the comparison of different regions and of different measured evapotranspiration values within a given region.

To obtain the highest possible yields of many agricultural crops, irrigation must be provided in an amount sufficient to prevent water from becoming a limiting factor. Knowledge of the potential evapotranspiration can therefore serve as a basis for planning the irrigation regime. In general, the actual evapotranspiration E_t from various crops will not equal the potential value E_{tp}, but in many cases, the maintenance of optimal soil-moisture conditions for maximal yields will result in E_t being nearly equal to, or a nearly constant fraction of, E_{tp}.

Various empirical approaches have been proposed for the estimation of potential evapotranspiration (e.g., Thornthwaite, 1948; Blaney and Criddle, 1950). The method proposed by Penman (1948) is physically based, and hence inherently more meaningful. His equation, based on a combination of the energy balance and aerodynamic transport considerations, is a major contribution in the field of agricultural and environmental physics.

The Dalton equation for evaporation from a saturated surface is

$$LE = (e_s - e)f(u) \tag{11.14}$$

where u is the windspeed above the surface, e_s is the saturated vapor pressure at the temperature of the surface, and e is the vapor pressure of the air above the surface.

If e_a is the saturated vapor pressure of the air, then

$$LE_a = (e_a - e)f(u) \tag{11.15}$$

and

$$\frac{E_a}{E} = 1 - \left[\frac{e_s - e_a}{e_s - e}\right] \tag{11.16}$$

is obtained by dividing equation (11.15) by (11.14) and rearranging of terms. Now, Penman assumed that $S = 0$ and he could write equation (11.13) as

$$\frac{J_n}{LE} = 1 + \beta \tag{11.17}$$

Since the Bowen ratio may be written as

$$\beta = \xi \left(\frac{T_s - T_a}{e_s - e}\right)$$

then

$$\frac{J_n}{LE} = 1 + \xi \left(\frac{T_s - T_a}{e_s - e} \right) \tag{11.18}$$

or by rearrangement we get

$$\frac{J_n}{LE} = 1 + \xi \left(\frac{e_s - e_a}{e_s - e} \right) \bigg/ \left(\frac{e_s - e_a}{T_s - T_a} \right) \tag{11.19}$$

In equation (11.19) we may write

$$\frac{e_s - e_a}{T_s - T_a} = \left(\frac{\Delta e}{\Delta T} \right)_{T \, = \, T_a} = \Delta$$

where Δ is the slope of the saturated vapor pressure-temperature curve. Now we may rewrite equation (11.19) as

$$\frac{J_n}{LE} = 1 + \frac{\xi}{\Delta} \left(\frac{e_s - e_a}{e_s - e} \right) \tag{11.20}$$

But since $(e_s - e_a)/(e_s - e) = 1 - E_a/E$ from equation (11.16) then by algebraic rearrangement we may write

$$LE = \frac{(\Delta/\xi)J_n + LE_a}{\Delta/\xi + 1} \tag{11.21}$$

Equation (11.21) is the Penman equation where

$LE_a = 0.35(e_a - e) \ (0.5 + U_2/100) \ \text{(mm/day)}$

$e_a = $ sat. vapor pressure at mean air temperature (mm Hg)

$e = $ mean vapor pressure in air

$U_2 = $ mean wind speed in miles/day at 2m above ground

This equation thus permits a calculation of the potential evapotranspiration rate from measurements of the net radiation, and of the temperature, vapor pressure, and wind velocity taken at one level above the field.

Other methods for the estimation of evapotranspiration from climatic data have been proposed by Thornthwaite (1948) and by Blaney and Criddle (1950). Actual evapotranspiration from an actively growing crop in the field generally constitutes a fraction (say, 50–100%) of the potential evapotranspiration as determined by the Penman equation or by evaporation pans (Fuchs, 1970).

I. Summary

The various soil-water flow processes, though differing in rate and direction, can be summarized in an overall equation called the field water balance, which itemizes all gains and losses of water to the system. A major item in this balance is the amount of water transmitted from the field to the atmosphere in the process of evapotranspiration, which in arid region commonly constitutes the principal consumer of water. The field water balance is intimately associated with the energy balance, which specifies how the supply of energy to the field (principally by radiation, and sometimes by advection as well) is partioned among the processes of evapotranspiration and of absorption by soil, air, and plants. The concept of potential evapotranspiration is an attempt to characterize the climatic environment in terms of its evaporative demand, or power. Actual evapotranspiration is generally appreciably lower than the potential evapotranspiration, owing to water supply limitations.

APPENDIX 1 *Numerical Solution of The Flow Equation*[1]

A. General

In essence, whenever we set forth to describe a phenomenon, we begin a model-building process. The model may be physical or it may be purely conceptual as is the hydrologic cycle discussed elsewhere in this book. When we construct a quantitative model, an example of which is the water-balance relationship also given elsewhere in this book, we write equations or construct graphs or do both.

Models rarely, if ever, duplicate their prototypes in every detail. This is particularly true when speaking of models of soil-water flow systems. Physical models may look like scaled-down versions of real systems, but the grain and pore sizes and their distributions within the porous media will not be scaled versions of their counterparts in the prototypes. Mathematical models of soil-water flow can usually be constructed only after making several simplifying assumptions, each of which detracts something from the ability of the model to portray the prototype. Nevertheless, careful model construction, whether physical or mathematical, can assure that the essence of the prototype is seen in the model. The point to recognize here is that a model user must view the data obtained from his model as approximate. One of

[1] This appendix is by C. R. Amerman, Research Hydraulic Engineer, USDA-ARS-SWCRD and Ass't. Professor of Soils, University of Wisconsin, Madison. Contribution from the Corn Belt Branch, Soil and Water Conservation Research Division, Agricultural Research Service, USDA, Madison, Wisconsin, in cooperation with the Wisconsin Agricultural Experiment Station.

his tasks is to determine the degree of approximation by whatever means possible.

Because the flow of water through soil is very complex and defies direct observation, one of the chief activities of many soil physicists is the building of conceptual, physical, or mathematical models of the phenomenon. In their case, the effort is not one simply of reproduction of a mechanism, but an integral part of the study of the mechanism.

In this appendix, we shall discuss one of the mathematical modeling techniques now being used—that of finite differences. This technique actually involves three modeling steps: (1) construct a mathematical model (a partial differential equation) of soil-water flow, (2) model this equation by means of a system of finite-difference equations, (3) model the finite-difference system by means of a digital computer program. Each of these steps introduces another order of approximation in the data finally obtained as representative of the real flow system under consideration.

B. The General Equation of Flow

The reader has already been introduced to quite a number of mathematical models dealing with the flow of water through soil. The basic model from which all the others derive is Eq. (5.7), reproduced here for convenience

$$\frac{\partial \theta}{\partial t} = \frac{\partial}{\partial x}\left(K\frac{\partial H_p}{\partial x}\right) + \frac{\partial}{\partial y}\left(K\frac{\partial H_p}{\partial y}\right) + \frac{\partial}{\partial z}\left(K\frac{\partial H_p}{\partial z}\right) + \frac{\partial K}{\partial z}$$

where θ is water content on volumetric basis; t time; K hydraulic conductivity, a function of water content, hence a function also of time and space; H_p soil-water pressure head (negative and designated ψ if θ is less than its value for saturation); and x, y, z distances along Cartesian coordinates.

We assume Eq. (5.7) to apply at every point where soil-water movement takes place, thus it may be considered to constitute the heart of the model. But to completely specify any model, we must describe its geometry. Further, by whatever type of model we simulate the equation, we must be able to specify conditions on the boundary either in terms of H_p itself or in terms of the gradient of H_p. If the flow condition is not steady, that is, if

$$\frac{\partial \theta}{\partial t} \neq 0$$

then we must also specify an initial condition, that is, the initial distribution of θ throughout the flow region. If the flow condition is steady, we need some assumed distribution of θ with which to begin computations. This assumed distribution may be completely arbitrary.

Equation (5.7) results from combining an equation of motion, Darcy's law, with a continuity equation and an equation of state. The assumptions underlying this development are: (1) the fluid of interest, water, is continuously connected throughout the flow region; (2) inertial forces are not significant as compared to viscous forces; (3) the fluid of interest, water, is incompressible.

In developing Eq. (5.7), several influences and mechanisms were not considered. To say that the above equation models soil-water flow, one must accept the following additional assumptions: (4) flow is isothermal; (5) the chemical nature of water does not change with time or position; (6) biological phenomena have no effect on soil-water flow; (7) air freely and instantaneously escapes from the system as water accumulates in it; (8) soil does not shrink or swell as water content changes.

The first three assumptions, those inherent to the development of Eq. (5.7), can all be shown to be reasonable, though the first may not be for very dry soil. Assumptions 5–8 are connected with mechanisms that we haven't yet learned to model successfully, so they must be accepted if we are to study soil-water movement at our present state of knowledge. Finally, we assume that the soil properties are: (9) homogeneous in space, and (10) isotropic; and we usually select initial conditions such that: (11) water content either increases or decreases monotonically, thus avoiding the effects of hysteresis of soil properties.

In constructing analytic models, such as Gardner's infiltration model, Eq. (6.4), assumptions 9–11 have always been made. It is conceivable, however, that if soil properties change in some smooth and regular manner in space, such variation could be modeled. If we wish to construct finite-difference models, we usually assume conditions 9–11, but there is no theoretical reason why we have to. There have, in fact, been finite-difference models which took account of hysteresis Rubin, 1967; Whisler and Klute, 1965). The reasons are practical—tremendous amounts of programming and computational time are needed to include nonhomogeneity in particular. Another factor is storage ("memory") limitations in computers. Non-homogeneity, if included in the model, would result in much larger storage requirements than a homogeneous model.

We wish to emphasize that all mathematical models derived from Eq. (5.7) or an equivalent form are restricted by assumptions 1–8. Most, either by necessity or convenience, are also restricted by assumptions 9–11.

Equation (5.7) is a second-order, partial differential equation of parabolic type. Since K is dependent upon θ, this equation is highly nonlinear. There is no way of solving it directly. The only approach is to make further simplifying assumptions or to approximate the equation with another equation that can be solved.

The first additional assumption usually made is that flow is one-dimensional and that it is either vertical or horizontal. In the former case, gravity effects must be considered; in the latter, gravity has no effect. The several equations for infiltration and redistribution elsewhere in this book are analytical models of the one-dimensional form of Eq. (5.7). Lately, investigators have also begun to consider the two-dimensional case.

We note, in passing, that analog models (resistance–capacitance networks or electrolytic solutions) of soil-water movement really are analog models of Eq. (5.7) or an equivalent form in one, two, or three space dimensions. Analogs then, besides being restricted by the assumptions or simplifications they themselves introduce, are restricted by the same basic assumptions as are mathematical models.

C. Finite-Differencing Concept

Finite differencing is a technique whereby continuous phenomena may be approximated by discrete functions. Rainfall, for example, is recorded as a continuous curve on the chart of a recording raingage. We determine the amount of rainfall accumulation $\mathbf{P}$ at any given time by direct reading of the chart. But in order to determine rainfall intensity $\mathbf{I}$, we must measure the rate of change of $\mathbf{P}$ with respect to time; that is

$$\mathbf{I} = \frac{d\mathbf{P}}{dt}$$

Since the rainfall accumulation trace is irregular, we cannot fit a curve of $\mathbf{P} = \mathbf{P}(t)$, so we cannot differentiate the relationship analytically; we have to do it by finite differences. At each point in time at which we want to know the intensity, we may select a short segment of the curve, Δt long and located so the point of interest is $\Delta t/2$ units of time from each end of the segment. Then, we determine $\mathbf{P}_1$ at the beginning of the segment and $\mathbf{P}_2$ at the end. The intensity is then

$$\mathbf{I} \approx \frac{\mathbf{P}_2 - \mathbf{P}_1}{\Delta t}$$

Provided that we can determine $\mathbf{P}$ with precision, the smaller we select Δt, the better is the approximation of $\mathbf{I}$.

The above is a simple illustration of the central concept of modeling partial differential equations by means of finite differences—each partial differential term can be replaced by a difference divided by a difference. In this manner, a partial differential equation can be converted to an algebraic equation involving values of the equation variables as they exist at certain discrete points in space and time.

One type of finite differencing, the explicit type, orders the discrete points and the differencing operations in such a manner that the resulting finite-difference equation contains only one unknown, and consequently, may be solved simply and directly. Another method, the implicit method, results in more than one unknown in each equation. In order to solve for the unknowns by this method, we must be able to develop a system of linear equations with at least as many equations as there are unknowns.

Finite-difference expressions pertain to specified discrete points in space and time and include small, absolutely specified distances in space and time. The discrete points are often called nodes and the intervals in space and time are called mesh increments. Taken together, the nodes and mesh increments form a solution mesh. In constructing a finite-difference model for a given soil-water flow situation, a solution mesh is devised so as to include the entire flow region. Usually, we try to simplify the geometry and select a mesh increment size such that each boundary will lie on a plane or line of nodes. This is not absolutely necessary, but the mechanics of solution are simplified a great deal if it can be done.

The mesh increment size in the solution mesh may or may not be equal in all spatial directions used, and they may or may not be uniform throughout the solution mesh. In either the temporal or the spatial portion of a solution mesh, smaller mesh increments may be used where the solution is expected to vary the most rapidly, while larger increments are used elsewhere.

A complete finite-difference model of a soil-water flow situation consists of a solution mesh, the finite-difference equation for each node in the mesh, a boundary condition specified at each boundary node, and an initial condition for unsteady flow or an assumed solution for steady flow at each node in that part of the mesh representing the flow system at the beginning of model operation.

Operation of the model consists of solving the finite-difference equations in some sequential manner so that the dependent variable is determined for each node of the solution mesh. If flow is unsteady, operation of the model consists of determining the distribution of the dependent variable over the entire spatial mesh for each time increment over some period of time specified by the operator.

The principal additional restriction introduced by a finite-difference model over those already introduced by the partial differential equation is that of accuracy, as compared with the solution of the partial differential equation itself. If a finite-difference model is valid, however, it converges to the partial differential equation as the mesh increments approach zero size. An estimate of accuracy may be obtained by operating the model for at least three different mesh increments and plotting the values of the dependent variable against mesh increment size. If the latter are very large, there will be no smooth curve; accuracy will be very poor and cannot be estimated. If too large a mesh increment is used, then the values of the dependent variable being determined by the model may oscillate, so that we may be able to detect this condition by considering the results of a single model run. If the mesh increments were small enough to obtain solutions of reasonable accuracy, then the graphs will be smooth curves converging to some constant value of the dependent variable. By extrapolation, we may determine the value for zero mesh-increment size and thus arrive at an estimate of the accuracy obtained by each mesh increment. We should then be able to select the proper mesh increment for future operation of the model so as to obtain solutions within some specified limit of accuracy.

A finite-difference model, being a discrete model, cannot completely fulfill the requirements of continuity. The operator may find, for instance, that when his model reaches a steady-state condition, the flux into and out of it may not be exactly equal. This does not invalidate the model, but is one of the restrictions on accuracy inherent in the model.

If we use an explicit finite-differencing technique, we must be careful that our model does not become unstable. Instability is indicated if the dependent variable oscillates with increasing amplitude as the solution progresses. This condition can be avoided by taking sufficiently small mesh increments. In unsteady soil-water flow problems, mesh increments for explicit formulations must be extremely small. This is a severe limitation, so we usually try to find implicit techniques.

Smith (1965) has written a highly readable text on the application of finite differencing to partial differential equations. Richtmyer and Morton (1967) discuss convergence and stability. If the reader wishes to apply finite differencing to more than one spatial dimension, he will need to familiarize himself with matrix algebra and its implementation on digital computers. Useful books on these subjects are Hadley (1961) and Varga (1962).

D. Finite-Difference Model of Steady, Saturated Flow

To illustrate the construction of a finite-difference model, we choose the case of vertical, steady-state, saturated flow through soil. The length of the flow region is L. This is a very simple situation that can be solved analytically.

Upper and lower boundary conditions must be specified. Let us say that we are supplying water at the soil surface at such a rate that a very thin film of water is maintained. Thus, the pressure on the surface is effectively zero. Let us also say that the discharge from the soil joins a water table at the same rate at which water is removed from beneath it, so that a pressure of zero is maintained at depth L in the soil. Since the pressure on both boundaries is zero, the pressure at every point in the soil is also zero and the potential gradient is therefore entirely gravitational and has the value unity. The potential at any point, then, is simply z, the elevation of that point with respect to some datum.

To model this situation numerically, we superimpose a finite-difference mesh on the length zero to L with a node at zero, a node at L and several nodes spaced equidistantly between them.

For saturated flow conditions, it is usually easiest to work directly with the potential form of Eq. (5.7):

$$\frac{\partial \theta}{\partial t} = \frac{\partial}{\partial x}\left(K \frac{\partial \phi}{\partial x}\right) + \frac{\partial}{\partial y}\left(K \frac{\partial \phi}{\partial y}\right) + \frac{\partial}{\partial z}\left(K \frac{\partial \phi}{\partial z}\right) \tag{1}$$

At saturation, moisture content does not change with time and K is constant if the soil is homogeneous and isotropic, as we assume it to be for the case at hand. For the one-dimensional case of saturated flow, then, Eq. (1) reduces to

$$\frac{\partial}{\partial x}\left(\frac{\partial \phi}{\partial x}\right) = 0 \tag{2}$$

To complete our model, we must write the finite-difference form for this equation at every node except the top and bottom nodes, where we know the pressures (and hence the potentials, since $\phi = z + H_p$).

To understand how Eq. (2) is finite-differenced, visualize three neighboring nodes and their two associated mesh increments. Assign the identifier 1 to the lower mesh increment and the identifier 2 to the upper. The identifiers $i-1, i$, and $i+1$ are assigned the upper, middle and lower nodes, respectively. Finite-differencing is accomplished as follows:

$$\frac{\partial}{\partial x}\left(\frac{\partial \phi}{\partial x}\right) \approx \frac{1}{\Delta x}\left(\left.\frac{\partial \phi}{\partial x}\right|_1 - \left.\frac{\partial \phi}{\partial x}\right|_2\right)$$

$$\left.\frac{\partial \phi}{\partial x}\right|_1 \approx \frac{\phi_{i+1} - \phi_i}{\Delta x}, \qquad \left.\frac{\partial \phi}{\partial x}\right|_2 \approx \frac{\phi_i - \phi_{i-1}}{\Delta x} \tag{3}$$

Thus,

$$\frac{\partial}{\partial x}\left(\frac{\partial \phi}{\partial x}\right) \approx \frac{\phi_{i+1} - 2\phi_i + \phi_{i-1}}{\Delta x^2} \tag{4}$$

and, substituting the right side of (4) into (2),

$$\phi_{i+1} - 2\phi_i + \phi_{i-1} = 0$$

or

$$\phi_i = \frac{\phi_{i+1} + \phi_{i-1}}{2} \tag{5}$$

The completed model consists of Eq. (5) for all interior nodes of the solution mesh and the two boundary conditions

$$\phi = z_0, \qquad x = 0$$

$$\phi = z_0 - L, \qquad x = L,$$

where z_0 is the equivalent of the soil surface above some datum.

Operation of the model is quite simple. Assume the position of the datum and establish the value z_0. Assume a starting value of ϕ at each interior node. We know from our previous analysis of this case that ϕ varies linearly from zero to L and that its value is equal to its elevation, so if the reader wishes to go through a trial operation of the model, he should select some different distribution. For example, set $\phi = 0$ at each interior node. Now, start at the first node beneath the surface and solve for ϕ_i, using z_0 for ϕ_{i-1}. The value of ϕ_i will be the first estimate of ϕ at that node. Next, consider the second node below the surface. In this case, the value to use for ϕ_{i-1} will be the value just computed for that node. Repeat this operation for each interior node in the solution mesh.

When the last node above the water table has been treated, the first iteration of the solution is finished. We now have a new approximation to the distribution of ϕ within our flow region. But it will not be very accurate, so we repeat the whole procedure as many times as necessary to reach a reasonable approximation to the solution. As the number of iterations increases, the difference between successive values of ϕ will become smaller. We may decide to accept the solution as being adequate if the differences between the ϕ values of succeeeding iterations are all less than 1% of the ϕ values.

This simple, even trivial, problem illustrates three points. The operation of a steady-state finite-difference model is characterized by (1) operator selection of some approximation to the unknown distribution of the dependent variable, (2) calculating by iteration from the assumed distribution to a closer approximation, where (3) the goodness of approximation to be attained is predetermined by the operator and is used to decide when to halt calculations.

If the reader has set up and operated the above-described model for a number of iterations, he will in his calculations have introduced another restriction into this model, that of roundoff error. For a simple problem of the type discussed, roundoff errors may not be serious. More-complex models usually are operated by means of computer programs, and, depending on the word length of the particular computer used, it may be necessary to use double-precision arithmetic to keep roundoff error within acceptable bounds.

Computations for steady-state or saturated flow models in one space dimension can be speeded up using a technique known as successive over-relaxation (SOR) and described by Smith (1965). In two space dimensions, fastest convergence of the solution is obtained using the alternating direction implicit (ADI) method. This method is discussed by Peaceman and Rachford (1955), Douglas and Peaceman (1955), and Amerman (1969). Taylor and Luthin (1963) have discussed a finite-difference model, using the SOR method, of a tile drainage system under steady flow conditions. In that paper, they also show how the model is implemented on a digital computer.

Before leaving saturated flow, we mention that Eq. (2) is often associated only with steady flow cases. In fact, the right-hand side is zero only because at saturation there is no further change in water content. Equation (2) is also applicable to cases where ϕ varies with time even though θ does not. In such a case, the change in the ϕ distribution with time is due to changes in the boundary conditions and a model of this situation must include a mechanism for applying these changes at the boundaries.

E. Finite-Difference Model of Unsteady, Unsaturated Flow

When unsteady, unsaturated flow must be modeled, time adds another dimension to the problem. We may continue to consider the flow region of one spatial dimension treated in the preceding section, with the water table maintained at depth L. Now, however, we start with an initially unsaturated soil above the water table. For simplicity, we wish to avoid hysteresis of soil properties. Therefore, a convenient initial condition for demonstration purposes is the distribution of potential that occurs when the soil is drained to static equilibrium. Static equilibrium occurs only where there is no potential gradient. Since $\phi = z + H_p$, zero potential gradient can occur only if the pressure gradient is equal in magnitude and opposite in sign to the gravity gradient. Since the gravity gradient is unity, then the pressure gradient is the negative of unity, and the pressure at a point in the soil is the negative of the elevation of that point above the water table.

To change the flow condition from one of static equilibrium, we shall at time zero introduce the same upper boundary condition as in the preceding section—a thin film of water on the surface so that the pressure on the upper boundary is zero. Since the water table continues at depth L, the lower boundary condition is also the same as in the preceding section.

The solution mesh at time zero and the boundary conditions during model operation, then, are the same as before. But, since we have added the time dimension, the complete solution mesh is now two-dimensional. A column in this mesh represents the soil profile at a given time. A row of nodes in the mesh represents a given soil depth through time.

To complete this model, we must again determine the finite-difference equation to be written for each node. One scheme for doing this was introduced into the literature pertaining to soil-water movement by Hanks and Bowers (1962), as they developed an implicit-type finite-difference model for infiltration into a layered soil. This scheme has been used by others in analyses involving one space dimension and serves as the basis for the alternating-direction implicit method as applied to situations involving two dimensions in space.

We start with the one-dimensional (vertical) form of Eq. (5.7), as before:

$$\frac{\partial \theta}{\partial t} = \frac{\partial}{\partial x}\left(K\frac{\partial H_p}{\partial x}\right) + \frac{\partial K}{\partial x} \tag{6}$$

This equation needs to be transformed so that we have only one dependent variable. Rubin (1966) discusses three possible ways of doing this. For this discussion, we choose the pressure head as the independent variable, which means that we introduce water capacity C into Eq. (6):

$$C = \frac{\partial \theta}{\partial H_p}$$

By the chain rule of calculus,

$$\frac{\partial \theta}{\partial t} = \frac{\partial \theta}{\partial H_p}\frac{\partial H_p}{\partial t} = C\frac{\partial H_p}{\partial t}$$

Finite-differencing involves subscripting and superscripting the dependent variable, and so we wish to simplify the symbols as much as possible. For the rest of this discussion, then, let

$$H = H_p = -\psi$$

where ψ was defined in Chapter 5 as the matric suction head. The form of the flow equation that we wish to model by finite differences is

$$C\frac{\partial H}{\partial t} = \frac{\partial}{\partial x}\left(K\frac{\partial H}{\partial x}\right) + \frac{\partial K}{\partial x} \tag{7}$$

Again assign the subscript identifier i to represent the depth of a node. A node i is midway between nodes $i-1$ above and $i+1$ below. Assign the superscript identifier j to represent time. A node j is preceded in time by node $j-1$ and followed in time by node $j+1$. Assign the identifier 1 to the mesh increment between nodes $i-1$ and i, and the identifier 2 to the mesh increment between nodes i and $i+1$. Thus,

$$C\frac{\partial H}{\partial t} \approx \bar{C}\frac{H_i^j - H_i^{j-1}}{\Delta t}$$

To represent $\bar{C}$, it is convenient to use the average of the C values corresponding to H values at nodes $(i, j - 1)$ and (i, j), i.e., the average of the C values at
time $t - 1$ and time t.

While operating the model so as to obtain a solution for a given time
increment, $t - 1$, i.e., column $j - 1$, represents the time at which H is known
at all points in the soil. Time t, column j, refers to the as yet unknown
distribution of H at the end of that time increment. Water capacity and
hydraulic conductivity, which, as we shall see, appears in the finite-difference
model in a form similar to that of C, are both dependent upon water content
(as also is H) and cannot be known *a priori* for time t. Thus, it is necessary to
assume values for C^j and K^j, calculate H^j, adjust C^j and K^j to correspond to
this estimate of H^j, then calculate a new estimate of H^j and so on until, as
with the steady-state solution, there is no more than some acceptable degree
of change in H^j from one iteration to the next. This iterative procedure of
calculating the distribution of H^j during a given time interval is one way of
linearizing Eq. (7). A reasonably fast convergence is obtained by making
the first estimates of C^j and K^j equal to C^{j-1} and K^{j-1}, respectively.

The first step in finite-differencing the first term on the right of Eq. (7)
is similar to that taken with the steady case,

$$\frac{\partial}{\partial x}\left(K\frac{\partial H}{\partial x}\right) \approx \frac{1}{\Delta x}\left(K\frac{\partial H}{\partial x}\bigg|_1 - K\frac{\partial H}{\partial x}\bigg|_2\right)$$

Since we are developing an implicit scheme, we want the finite-difference
form of the above to include H^j as well as H^{j-1}. One way to do this is

$$K\frac{\partial h}{\partial x}\bigg|_1 \approx \frac{\bar{K}_1}{\Delta x}\left(\frac{H_{i-1}^{j-1} + H_{i-1}^{j}}{2} - \frac{H_i^{j-1} + H_i^{j}}{2}\right)$$

and

$$K\frac{\partial h}{\partial x}\bigg|_2 \approx \frac{\bar{K}_2}{\Delta x}\left(\frac{H_i^{j-1} + H_i^{j}}{2} - \frac{H_{i+1}^{j-1} + H_{i+1}^{j}}{2}\right)$$

where $\bar{K}_1$ is the average of the K values corresponding to the H values at
nodes $(i - 1, j - 1)$, $(i - 1, j)$, $(i, j - 1)$, and (i, j), and $\bar{K}_2$ is similarly associated with nodes $(i, j - 1)$, (i, j) $(i + 1, j - 1)$, and $(i + 1, j)$. Completing
the finite-difference form of the first term on the right of Eq. (7), we have

$$\frac{\partial}{\partial x}\left(K\frac{\partial H}{\partial x}\right) \approx \frac{\bar{K}_1}{2(\Delta x)^2}(H_{i-1}^{j-1} + H_{i-1}^{j} - H_i^{j-1} - H_i^{j})$$

$$-\frac{\bar{K}_2}{2(\Delta x)^2}(H_i^{j-1} + H_i^{j} - H_{i-1}^{j-1} - H_{i+1}^{j})$$

Finally, we choose to finite-difference the last term of Eq. (7) as

$$\frac{\partial K}{\partial x} \approx \frac{\overline{K}_1 - \overline{K}_2}{\Delta x}$$

By substituting the several finite-difference forms into Eq. (7), rearranging and gathering unknowns on the left side of the equation, we obtain

$$-\overline{K}_1 H_{i-1}^j + \left(\frac{2(\Delta x)^2 \overline{C}}{\Delta t} + \overline{K}_1 + \overline{K}_2\right) H_i^j - \overline{K}_2 H_{i+1}^j$$

$$= \overline{K}_1 H_{i-1}^{j-1} + \left(\frac{2(\Delta x)^2 \overline{C}}{\Delta t} - \overline{K}_1 - \overline{K}_2\right) H_i^{j-1} + \overline{K}_2 H_{i+1}^{j-1}$$

$$+ 2(\Delta x)(\overline{K}_1 - \overline{K}_2) \tag{8}$$

We complete our model, then, by writing this linear equation (8) for each interior node of the solution mesh. But each equation except the one at the first node below the surface and the first node above the water table contains three unknowns, the quantities H^j. The exceptions each contain two unknowns; by reason of the boundary conditions, at the node beneath the surface and at the node above the water table, we know H_{i-1}^j and H_{i+1}^j, respectively. Thus, we cannot solve each equation explicitly, but must solve a system of simultaneous equations. Considering the set of equations for a single column, say j, of our solution mesh, we can determine that if there are N interior nodes, then there are N equations and N unknowns. So we shall operate the model by simultaneously calculating the H distribution for all depths over a single time increment, including several iterations to remove nonlinearity due to the dependence of C and K on moisture content, and then move to the next time increment and repeat the operations. In this way, we shall march our solution out through time, until we reach a predetermined condition such as the steady flow condition considered in the previous section.

The coefficients of the set of simultaneous equations for a given time increment form a tridiagonal matrix. For this type of matrix, the technique of matrix inversion known as Gaussian elimination can be set up as a simple, rapidly accomplished recursive computation, discussed by Richtmyer and Morton (1967), Peaceman and Rachford (1965), and Amerman (1969).

In describing the technique, it is convenient to note that each coefficient of an H term on the left of Eq. (8) is known, that the entire right side is known, and that, since we are considering a system of equations pertaining to only one time interval, we may dispense with superscripts. Thus, Eq. (8) can be written in the form

$$A_i H_{i-1} + B_i H_i + C_i H_{i+1} = D_i \tag{9}$$

A, B, C, and D are known by assumption or by corrected assumption for each iteration, but must be changed between the iterations required to accomplish the solution of a time step. The recursive formula by which H may be calculated is

$$H_i = g_i - b_i H_{i+1} \tag{10}$$

where, for all interior nodes,

$$g_i = \frac{D_i - A_i g_{i-1}}{B_i - A_i b_{i-1}} \quad \text{and} \quad b_i = \frac{C_i}{B_i - A_i b_{i-1}}$$

In applying this technique, we begin at one end of the jth column of nodes and sweep to the other, calculating the g and b values. Then, we sweep the column in the other direction, calculating the H values. To obtain starting values of b and g, write Eq. (9) for the first node ($i = 1$) below the surface,

$$A_1 H_0 + B_1 H_1 + C_1 H_2 = D_1$$

in which H_0, the surface boundary condition, is known. Rearranging,

$$H_1 = \frac{D_1 - A_1 H_0}{B_1} - \frac{C_1}{B_1} H_2 \tag{11}$$

Comparing the coefficients of Eq. (11) with those of Eq. (10), we recognize that

$$g_1 = \frac{D_1 - A_1 H_0}{B_1}, \quad b_1 = \frac{C_1}{B_1}$$

The above-described method is implicit and, according to Richtmyer and Morton (1967), is unconditionally stable where C and K are constants. Stability with nonconstant K and C has not been analyzed, but numerous applications of this method have not raised any problems in that regard. Choice of mesh increment lengths Δx and Δt depend entirely on the degree of accuracy needed. As the solution progresses through time, the H distribution changes more and more slowly, so that it is possible to lengthen Δt without losing accuracy.

Finite-difference models of unsteady flow regions in two space dimensions have been constructed (Amerman, 1969; Hornberger et al., 1969; Rubin, 1968; Taylor and Luthin, 1969) and can be used. If certain gaps in the mathematics of finite differencing can be filled, however, such models will be much improved.

F. Summary

One way to model the flow of water through soil is by means of a numerical technique known as finite differencing. Such models are based on the partial differential equation of flow, so are subject to all the assumptions and restrictions introduced by that equation. Finite differencing adds another degree of approximation because it is a discrete model of a continuous phenomenon. In constructing a finite-difference model, we change our mathematical model from one involving a nonlinear partial differential equation which cannot be solved directly to another mathematical model which is either an algebraic equation or a linear system of algebraic equations. Conceptually, a finite-difference model is built upon a solution mesh consisting of discrete nodes separated by discrete space and time intervals and superimposed over the soil-water flow region. Mathematically, the model consists of an algebraic equation for each node, of *a priori* known initial conditions at each node in the spatial mesh for unsteady problems or arbitrarily assumed starting conditions at each node for steady problems, and of boundary conditions, known throughout time, at each boundary node. These models must usually be operated on a digital computer.

Finite-difference models have the advantage that experimentally determined curves relating pressure head, water capacity, and hydraulic conductivity to water content can be used directly in the model. There is no need to fit an equation to such curves. These models also can be constructed for flow regions of nonregular geometry and for flow regions in which soil properties can vary in any manner from one point to another.

APPENDIX **2** *Mathematical Outline of Variables-Separable (Boltzmann) Solution of The Flow Equation*[1]

Solution of the general flow equation for unsaturated soils can be made simpler by first applying the Kirchoff change of variables to replace θ by a more easily manipulated quantity φ, defined as

$$\varphi = \int_{\psi}^{o} K(\psi)\, d\psi = \int_{\theta}^{\theta \text{sat}} D(\theta)\, d\theta \qquad (1)$$

so that[2]

$$\frac{-d\varphi}{d\psi} = D(\theta)\frac{d\theta}{d\psi} = K(\psi) \qquad (2)$$

This alteration has the advantage that $\nabla\varphi = -K\nabla\psi = -D\nabla\theta$, which has the effect of putting the K and D functions outside the operators in the flow equation.

However, since the interpretation of φ is slightly more difficult than the interpretation of θ, both forms of the manipulation will be indicated below.

---

[1] This appendix is by E. E. Miller, Department of Soil Science and Department of Physics, The University of Wisconsin, Madison, Wisconsin.

[2] Here $K(\psi)$ is the hydraulic conductivity, K, a function of the matric suction head, ψ; and $D(\theta)$ is the diffusivity, D, a function of the volumetric wetness, θ.

FOR ONE-DIMENSIONAL, MONOTONIC, GRAVITY FREE FLOW

Original form: Kirchoff form:

$$\frac{\partial \theta}{\partial t} = \nabla[D(\theta)\,\nabla\theta] \quad (3a) \qquad \frac{\partial \varphi}{\partial t} = D(\varphi)\,\nabla^2 \varphi \quad (3b)$$

(The simpler solution for φ can be read off from the solution for θ below by changing all small θ symbols to small φ symbols, all cap Θ symbols to cap Φ symbols, and omitting all the terms in boxes.) Subscripts will be used to abbreviate partial derivatives, and primes on a function symbol denote differentiation with respect to its argument.

All three systems of coordinates can be handled at once with the short-hand of a universal spatial coordinate, x, that represents radius, r, in cylindrical and spherical coordinates; along with an integer n, which assumes the values 0 for Cartesian, 1 for cylindrical, and 2 for spherical coordinates. In this telescoped notation, Eqs. (3a) and (3b) can be written for all three coordinate systems at once in the form

$$\theta_t = Dx^{-n}\,\partial/\partial x\,(x^n \cdot \theta_x) + \boxed{D'\theta_x{}^2} \tag{4}$$

Now we want to convert (4) into a form in which the coordinate x appears in the role of independent variable. To this end we must use some differential calculus: (1) Maxwell's relation, $l_m m_n n_l = -1$; (2) from differentiation of $l_m = (m_l)^{-1}$ the relation $l_{mm} = -m_{ll}/(m_l)^3$. Employing these relations we convert (4) to

$$x_t = D[x_{\theta\theta}/(x_\theta)^2 - n/x] - \boxed{D'/x_\theta} \tag{5}$$

In this form we are able to apply the variables-separable technique, assuming that x is a function Θ of θ alone times a function T of t alone, and finding that this assumption does satisfy (5).

Substituting

$$x = \Theta\,(\theta) \cdot T(t) \tag{6}$$

into (5) we get,

$$TT' = (D\Theta)[(\Theta''/\Theta')^2 - n/\Theta] - \boxed{D'/\Theta\Theta'} = \text{constant} \tag{7}$$

This is the normal result in the variables-separable technique in which a function of one variable alone is equal to a function of the other variable alone which holds if they are equal to a constant.

If we choose the constant to be $M^2/2$, we get $2TT' = (d/dt)(T^2) = M^2$. Integrating[3] and taking the square root, we obtain

$$T = Mt^{1/2} \tag{8}$$

which is handy. Putting (8) into (6), we get $x = t^{1/2} \cdot M \cdot \Theta(\theta)$ which suggests defining a function $B(\theta)$ so that

$$x/t^{1/2} = B(\theta) \tag{9}$$

i.e., $B(\theta)/M = \Theta(\theta)$ which can be substituted into (7) to give

$$1 = \frac{2D(\theta)}{B(\theta)}\left[\frac{B''(\theta)}{[B'(\theta)]^2} - \frac{n}{B(\theta)}\right] - \boxed{\frac{2D'(\theta)}{B(\theta)B'(\theta)}} \tag{10}$$

If it is desired to solve this for $B''(\theta)$ we get the form

$$B'' = (B')^2[B/2D + n/B] + \boxed{B'\, D'/D} \tag{11}$$

Of course we now have (10) and (11) as *ordinary* differential equations on $B(\theta)$, involving the specified function $D(\theta)$.

NOTES

Coordinates: Cartesian, $n = 0$; cylindrical, $n = 1$; spherical, $n = 2$.

Interpretation: B represents the Boltzmann group $(x/\sqrt{t})$ which is just a function of θ (or φ, omitting box) satisfying the above ordinary differential equation.

Parametric families: The solution of Eq. (11) is a two-parameter family. To see this conceptionally, imagine that for a given n, one could, at a given θ (or φ), choose both B and B' arbitrarily; compute B'' from (11); then extend B and B' a small distance $\pm \Delta\theta$ (or $\Delta\varphi$), and repeat in the usual simple fashion to develop a solution. Since B and B' were chosen arbitrarily, they represent two separate parameters for *two* families of solutions generated. In the Cartesian case, for example, these two parameters can be physically correlated with the initial and the face-applied values of suction.

Cartesian θ-dependent version: The final equation above with $n = 0$ can be rearranged into the form

$$\left(\frac{B}{2}\right)\frac{1}{B'} = \frac{1}{B'}\left(\frac{DB''}{(B')^2} - \frac{D'}{B'}\right) \tag{12}$$

which in turn can be rewritten as Eq. (5.21) of section F, Chapter 5.

---

[3] The boundary condition on the system operating on equation (6) requires that $T \to 0$ as $t \to 0$ so no constant $T_0{}^2$ appears in Eq. (8).

A word should be added about boundary conditions. In the foregoing discussion we have mentioned several related families of solutions for the differential equation. Of these solutions, only those which happen to fit a set of boundary conditions that can be realized in practice are of particular interest to us. For example, the one-dimensional solutions described for cylindrical and spherical coordinates all exhibit a pole (an infinite value) at the origin ($r = 0$). Although it is true that such solutions could be tested experimentally (by fitting them with time-dependent boundary conditions precomputed for some finite $r =$ constant surface) such artificially contrived tests are of limited practical interest. In contrast, the Cartesian solution is not only easy to test experimentally, but it constitutes a very useful early-time approximation to many transient processes that occur in practice. (For example, in the earliest stages of vertical infiltration, the suction gradients completely negate the effect of gravity so that this solution is a good approximation.)

References

Alway, F. J., and McDole, G. R. (1917). Relation of the water-retaining capacity of a soil to its hygroscopic coefficient. *J. Agr. Res.* **9**, 27–71.

Amerman, C. R. (1969). Finite difference solutions of unsteady, two-dimensional, partially saturated porous media flow. PhD Thesis, Purdue Univ. Lafayette, Indiana.

Anderson, M. S. (1926). Properties of soil colloidal material. *U. S. Dept. Agr. Bull.* 1452.

Aslyng, H. C. *et al.* (1962). Soil physics terminology, draft report. Int. Soc. of Soil Sci. Bull. 20.

Aslyng, H. C. *et al.* (1963). Soil physics terminology. Inter. Soc. of Soil Sci. Bull. 23, p.7.

Aylmore, L. A. O., and Quirk, J. P. (1959). Swelling of claywater systems. *Nature* **183**, 1752.

Babcock, K. L. (1963). Theory of the chemical properties of soil colloidal systems at equilibrium. *Hilgardia* **34**, 417–542.

Babcock, K. L., and Overstreet, R. (1955). The thermodynamics of soil moisture: a new approach. *Soil Sci.* **80**, 257–263.

Babcock, K. L., and Overstreet, R. (1957). The extra-thermodynamics of soil moisture. *Soil Sci.* **83**, 455–464.

Barber, S. A. (1962). A diffusion and mass-flow concept of soil nutrient availability. *Soil Sci.* **93**, 39–49

Barrow, G. M. (1961). " Physical chemistry." McGraw-Hill, New York.

Baver, L. D. (1956). "Soil Physics." Wiley, New York.

Baver, L. D., and Farnsworth, R. B. (1940). Soil structure effects in the growth of sugar beets. *Soil Sci. Soc. Amer. Proc.* **5**, 45–48.

Bear, F. E. (1955). "Chemistry of Soil." Reinhold, New York.

Bear, J., Zaslavsky, D., and Irmay, S. (1968). "Physical Principles of Water Percolation and Seepage." UNESCO, Paris.

Bertrand, A. R. (1967). Water conservation through improved practices. *In* " Plant Environment and Efficient Water Use." Amer. Soc. Agron.

Black, C. A. ed. (1965). "Methods of Soil Analysis." Amer. Soc. Agron. Monograph 9.

Black, C. A. (1968). "Soil Plant Relationships." Wiley, New York.

Black, T. A., Thurtell, G. W., and Tanner, C. B. (1968). Hydraulic load-cell lysimeter, construction, calibration, and tests. *Soil Sci. Soc. Amer. Proc.* **32**, 623–629.

Black, T. A., Gardner, W. R., and Thurtell, G. W. (1969). The prediction of evaporation, drainage and soil water storage for a bare soil. *Soil Sci. Soc. Amer. Proc.* **33**, 655–660.

Blaney, H. F., and Criddle, W. D. (1950). Determining water requirements in irrigated areas from climatological and irrigation data. U.S. Soil Conser. Serv. Tech. Publ. 96.

Bodman, G. B., and Coleman, E. A. (1944). Moisture and energy conditions during downward entry of water into soils. *Soil Sci. Soc. Amer. Proc.* **8**, 116–122.

Boekel, P. (1963). Soil structure and plant growth. *Neth. J. Agr. Sci.* **11**, 120–127.

Boersma, L. (1965a). Field measurement of hydraulic conductivity below a water table, In "Methods of Soil Analysis," p. 222–223. Amer. Soc. Agron. Monograph 9.

Boersma, L. (1965b). Field measurement of hydraulic conductivity above a water table. In "Methods of Soil Analysis," p. 234–252. Amer. Soc. Agron. Monograph 9.

Bolt, G. H., and Frissel, M. J. (1960). Thermodynamics of soil moisture. *Neth. J. Agr. Sci.* **8**, 57–78.

Bomba, S. J. (1968). Hysteresis and time-scale invariance in a glass-lead medium. PhD Thesis, Univ. Wisconsin, Madison, Wisconsin.

Bond, J. J., and Willis, W. O. (1969). Soil water evaporations: surface residue rate and placement effects. *Soil Sci. Amer. Proc.* **33**, 445–448.

Bower, C. A., and Goertzen, J. O. (1959). Surface area of soils and clays by an equilibrium ethylene glycol method. *Soil Sci.* **87**, 289–292.

Bower, C. A., Gardner, W. R., and Goertzen, J. O. (1957). Dynamics of cation exchange in soil columns. *Soil Sci. Soc. Amer. Proc.* **21**, 20–24.

Bouwer, H. (1959). Theoretical aspects of flow above the water table in tile drainage of shallow homogeneous soil. *Soil Sci. Soc. Amer. Proc.* **23**, 200–203.

Bouwer, H. (1961). A double tube method for measuring hydraulic conductivity of soil in sites above a water table. *Soil Sci. Amer. Proc.* **25**, 334–342.

Bouwer, H. (1962a). Analyzing groundwater mounds by resistance network analog. *J. Irrig. Dram. Div., Proc. Amer. Soc. Civ. Eng.* **88**, IR 3, 15–36.

Bouwer, H. (1962b). Field determination of hydraulic conductivity above a water table with the double tube method. *Soil Sci. Soc. Amer. Proc.* **26**, 330–335.

Bouwer, H. (1964). Resistance network analogs for solving ground-water problems. *Ground-Water* 2#(3), 1–7.

Bouwer, H., and van Schilfgaarde, J. (1963). Simplified method of predicting fall of water table in drained land. *Trans. Amer. Soc. Agr. Eng.* **6**, 288–291; 296.

Bouyoucos, G. J. (1915). Effect of temperature on the movement of water vapor and capillary moisture in soils. *J. Agr. Res.* **5**, 141–172.

Bouyoucos, G. J., and Mick, A. H. (1940). An electrical resistance method for the continuous measurement of soil moisture under field conditions. Michigan Agr. Exp. Sta. Tech. Bull. 172.

Bray, R. H. (1954). Nutrient mobility concept of soil-plant relationships. *Soil Sci.* **78**, 9–22.

Bresler, E. (1967). A model for tracing salt distribution in the soil profile and estimating the efficient combinations of water quality and quantity. *Soil Sci.* **104**, 227–233.

Bresler, E., and Hanks, R. J. (1969). Numerical method for estimating simultaneous flow of water and salt in unsaturated soil. *Soil Sci. Soc. Amer. Proc.* *33*, 827–832.

Bresler, E. (1970). Solute movement in soils. *In* "Irrigation in Arid Zones," pp. 109–119. Volcanic Institute of Agricultural Research, Rehovot, Israel.

Briggs, L. J., and Shantz, H. L. (1912). The relative wilting coefficient for different plants. *Bot. Gaz. (Chicago)* **53**, 229–235.

Brooks, R. H., Bower, C. A., and Reeve, R. C. (1956). The effect of various exchangeable cations upon the physical condition of soils. *Soil Sci. Soc. Amer. Proc.* **20**, 325–327.

Bruce, R. R., and Klute, A. (1956). The measurement of soil water diffusivity. *Soil Sci. Soc. Amer. Proc.* **20**, 458–462.

Brunauer, S., Emmett, P. H., and Teller, E. (1938). Adsorption of gases in multimolecular layers. *J. Amer. Chem. Soc.* **60**, 309–319.

Buckingham, E. (1907). U.S. Dept. of Agr. Bur. of Soils, Bull. 38.

Buckman, H. O., and Brady, N. C. (1960). "The Nature and Properties of Soils." Macmillan, New York.

Burwell, R. E., and Larson, W. E. (1969). Infiltration as influenced by tillage-induced random roughness and pore space. *Soil Sci. Amer. Proc.* **33**, 449–452.

Buswell, A. M., and Rodebush, W. H. (1956). Water. *Sci. Amer.* **202**, 1–10.

Carman, P. C. (1939). *J. Agr. Sci.* **29**, 262.

Carslaw, H. S., and Jaeger, J. C. (1959). "Conduction of Heat in Solids." Oxford Univ. Press, London and New York.

Cary, J. W. (1963). Onsager's relations and the non-isothermal diffusion of water vapor. *J. Phys. Chem.* **67**, 126–129.

Cary, J. W. (1964). An evaporation experiment and its irreversible thermodynamics. *Int. J. Heat Mass Transfer* **7**, 531–538.

Cary, J. W. (1965). Waterflux in moist soil: Thermal versus suction gradients. *Soil Sci.* **100**, 168–175.

Cary, J. W. (1966). Soil moisture transport due to thermal gradients: Practical aspects. *Soil Sci. Soc. Amer. Proc.* **30**, 428–433.

Cary, J. W. (1968). An instrument for *in situ* measurement of soil moisture flow and suction. *Soil Sci. Soc. Amer. Proc.* **32**, 3–5.

Cary, J. W., and Taylor, S. A. (1962). The interaction of the simultaneous diffusions of heat and water vapor. *Soil Sci. Soc. Amer. Proc.* **26**, 413–416.

Casagrande, A. (1937). Seepage through dams. *J. New England Water Works Ass.* **51**, 131–172.

Childs, E. C. (1940). The use of soil moisture characteristics in soil studies. *Soil Sci.* **50**, 239–252.

Childs, E. C. (1947). The water table equipotentials and streamlines in drained land. *Soil Sci.* **63**, 361–376.

Childs, E. C. (1969). "An Introduction to the Physical Basis of Soil Water Phenomena." Wiley (Interscience), New York.

Childs, E. C., and Collis-George, N. (1950). The permeability of porous materials. *Proc. Roy. Soc.* **201A**, 392–405.

Childs, E. C., and Poulovassilis, A. (1962). The moisture profile above a moving water table. *J. Soil Sci.* **13**, 272–285.

Collis-George, N., and Youngs, E. G. (1958). Some factors determining water table heights in drained homogeneous soils. *J. Soil Sci.* **9**, 332–338.

Colman, E. A., and Bodman, G. B. (1945). Moisture and energy conditions during downward entry of water into moist and layered soils. *Soil Sci. Soc. Amer. Proc.* **9**, 3–11.

Colman, E. A., and Hendrix, T. M. (1949). Fiberglass electrical soil moisture instrument. *Soil Sci.* **67**, 425–438.

Covey, W. (1963). Mathematical study of the first stage of drying of a moist soil. *Soil Sci. Soc. Amer. Proc.* **27**, 130–134.

Cowan, I. R. (1965). Transport of water in soil-plant-atmosphere system. *J. Appl. Ecol.* **2**, 221–229.

Crank, J. (1956). "The Mathematics of Diffusion." Oxford Univ. Press, London, and New York.

Currie, J. A. (1961). Gaseous diffusion in porous media Part 3—Wet granular material. *Brit. J. Appl. Phys.* **12**, 275–281.

Dalton, F. N., and Rawlins, S. L. (1968). Design criteria for peltier effect thermocouple psychrometers. *Soil Sci.* **105**, 12–17.

Darcy, Henry (1856). "Les Fontaines Publique de la Ville de Dijon." Dalmont, Paris.

Day, P. R. (1956). Dispersion of a moving salt-water boundary advancing through saturated sand. *Trans. Amer. Geophys. Union* **37**, 595–601.

Day, P. R. (1965). Particle fractionation and particle size analysis. *Agronomy* **9**, 545–567.

Day, P. R., and Luthin, J. (1956). A numerical solution of the differential equation of flow for a vertical drainage problem. *Soil Sci. Soc. Amer. Proc.* **20**, 443–447.

Day, P. R., and Forsythe, W. M. (1957). Hydro-dynamic dispersion of solutes in the soil moisture stream. *Soil Sci. Soc. Amer. Proc.* **21**, 477–480.

Deacon, E. L., Priestley, C. H. B., and Swinbank, W. C. (1958). Evaporation and the water balance. *Climatol.-Rev. Res. Arid Zone Res.* 9–34 (UNESCO).

de Boer, J. H. (1953). "The Dynamical Character of Adsorption." Oxford Univ. Press, London and New York.

Decker, W. L. (1959). Variations in the net exchange of radiation from vegetation of different heights. *J. Geophys. Res.* **64**, 1617–1619.

de Groot, S. R. (1963). "Thermodynamics of Irreversible Processes." North-Holland, Amsterdam.

de Jong, E. (1968). Applications of thermodynamics to soil moisture. *Proc. Hydrology Symp. 6th*, pp. 25–48. National Research Council of Canada.

de Wit, C. T. (1958). Transpiration and crop yields. *Versl. Landbovwk. Onderz.* No. 646, p. 88.

Denmead, O. T., and Shaw, R. H. (1962). Availability of soil water to plants as affected by soil moisture content and meteorological conditions. *Agron. J.* **54**, 385–390.

Deryaguin, B. V., and Melnikova, M. K. (1958). Mechanism of moisture equilibrium and migration in soils. Water and its conduction in soils. Int. Symp. Highway Res. Board, Spec. Rep. 40, pp. 43–54.

Deryaguin, B. V., Talaev, M. V., and Fedyakin, N. N. (1965). *Dokl. Akad. Nauk. SSSR* **165**, 597 [English transl.: *Proc. Acad. Sci. USSR Phys. Chem.* **165**, 807].

Deshpande, T. L., Greenland, D. J., and Quirk, J. P. (1964). Role of iron oxides in the bonding of soil particles. *Nature* **201**, 107–108.

de Vries, J., and King, K. M. (1961). Note on the volume of influence of a neutron surface moisture probe. *Can. J. Soil Sci.* **41**, 253–257.

Donnan, W. W. (1947). Model tests of a tile-spacing formula. *Soil Sci. Amer. Proc.* **11**, 131–136.

Douglas, J. Jr., and Peaceman, D. W. (1955). Numerical solution of two-dimensional heat flow problems. *Amer. Inst. Chem. Eng.* **1**, 505–512.

Dyal, R. S., and Hendricks, S. B. (1950). Total surface of clays in polar liquids as a characteristic index. *Soil Sci.* **69**, 421–432.

Edlefsen, N. E., and Anderson, A. B. C. (1943). Thermodynamics of soil moisture. *Hilgardia* **15**, 31–298.

Edwards, W. M., and Larson, W. E. (1970). Infiltration of water into soils as influenced by surface seal development. *Soil Sci. Soc. Amer. Proc.* (in press).

Ekern, P. C. (1950). Raindrop impact as the force initiating soil corosion. *Soil Sci. Soc. Amer. Proc.* **15**, 7–10.

Emerson, W. W. (1959). The structure of soil crumbs. *J. Soil Sci.* **10**, 235.

Erickson, A. E., and van Doren, D. M. (1960). The relation of plant growth and yield to oxygen availability. *Trans. Intern. Congr. Soil Sci., Madison 7th* **3**, 428–434.

Evans, D. D. (1965). Gas movement. *In* "Methods of soil analysis, pp. 319–330. Amer. Soc. Agron. Monograph 9.

Evans, L. T., and Russell, E. W. (1959). The adsorption of humic and fulvic acids by clays. *J. Soil Sci.* **10**, 119–132.

Ferguson, H., and Gardner, W. H. (1963). Diffusion theory applied to water flow data using gamma ray absorption. *Soil Sci. Soc. Amer. Proc.* **27**, 243–245.

Fireman, M. (1957). Salinity and alkali problems in relation to high water tables in soils. *In* "Drainage of Agricultural Lands," pp. 505–513. Amer. Soc. Agron. Monograph 7.

Fisher, E. A. (1923). Some factors affecting the evaporation of water from soil. *J. Agr. Sci.* **13**, 121–143.

Forchheimer, P. (1930). "Hydraulik," 3rd ed. Teubner, Leipzig and Berlin.

Forsgate, J. A., Hosegood, P. H., and McCulloch, J. S. G. (1965). Design and installation of semienclosed hydraulic lysimeters. *Agr. Meteorol.* **2**, 43–52.

Frank, H. S., and Wen, W. (1957). Structural aspects of ion-solvent interaction in aqueous solutions: a suggested picture of water structure. *Discuss. Faraday Soc.* **24**, 133–140.

Franzini, J. B. (1951). *Trans. Amer. Geophys. Union* **32**, 443.

Fritton, D. D., Kirkham, D., and Shaw, R. H. (1970). Soil water evaporation, isothermal diffusion, and heat and water transfer. *Soil Sci. Soc. Amer. Proc. 34*, 183–189.

Fuchs, M. (1970). Evapotranspiration, *In* "Irrigation in Arid Zones," pp. 155–162. Volcanic Institute of Agricultural Research, Rehovot, Israel.

Fuchs, M., Tanner, C. B., Thurtell, G. W., and Black, T. A. (1969). Evaporation from drying surfaces by the combination method. *Agron. J.* **61**, 22–26.

Gardner, W. (1920). The capillary potential and its relation to soil moisture constants. *Soil Sci.* **10**, 357–359.

Gardner, W. H. (1965). Water content. *In* "Methods of Soil Analysis," pp. 82–127. Amer. Soc. Agron. Monograph 9.

Gardner, W. R. (1956). Calculation of capillary conductivity from pressure plate ouflow data. *Soil Sci. Soc. Amer. Proc.* **20**, 317–320.

Gardner, W. R. (1958). Some steady state solutions of the unsaturated moisture flow equation with application to evaporation from a water table. *Soil Sci.* **85**, 228–232.

Gardner, W. R. (1959). Solutions of the flow equation for the drying of soils and other porous media. *Soil Sci. Soc. Amer. Proc.* **23**, 183–187.

Gardner, W. R. (1960a). Dynamic aspects of water availability to plants. *Soil Sci.* **89**, 63–73.

Gardner, W. R. (1960b). Soil water relations in arid and semi-arid conditions. Plant water relationships in arid and semiarid conditions. UNESCO **15**, 37–61.

Gardner, W. R. (1962a). Approximate solution of a non-steady state drainage problem. *Soil Sci. Soc. Amer. Proc.* **26**, 129–132.

Gardner, W. R. (1962b). Note on the separation and solution of diffusion type equations. *Soil Sci. Soc. Amer. Proc.* **26**, 404.

Gardner, W. R. (1964). Relation of root distribution to water uptake and availability. *Agron. J.* **56**: 35–41.

Gardner, W. R. (1965). Movement of nitrogen in soil. *In* "Soil Nitrogen," pp. 555–572. Amer. Soc. Agron. Monograph 10.

Gardner, W. R. (1968). Availability and measurement of soil water. *In* "Water Deficits and Plant Growth," Vol. 1, pp. 107–135. Academic Press, New York.

Gardner, W. R. (1970). Field measurement of soil water diffusivity. *Soil Sci. Soc. Amer. Proc.* (in press).

Gardner, W. R., and Kirkham, D. (1952). Determination of soil moisture by neutron scattering. *Soil Sci.* **73**, 391–401.

Gardner, W. R., and Brooks, R. H. (1956). A descriptive theory of leaching. *Soil Sci.* **83**, 295–304.

Gardner, W. R., and Fireman, M. (1958). Laboratory studies of evaporation from soil columns in the presence of a water table. *Soil Sci.* **85**, 244–249.

Gardner, W. R., and Mayhugh, M. S. (1958). Solutions and tests of the diffusion equation for the movement of water in soil. *Soil Sci. Soc. Amer. Proc.* **22**, 197–201.

Gardner, W. R., and Hillel, D. I. (1962). The relation of external evaporative conditions to the drying of soils. *J. Geophys. Res.* **67**, 4319–4325.

Gardner, W. R., and Ehlig, C. F. (1962). Some observations on the movement of water to plant roots. *Agron. J.* **54**, 453–456.

Gardner, W. R. and Ehlig, C. F. (1963). The influence of soil water on transpiration by plants. *J. Geophys. Res. 68*, 5719–5724.

Gardner, W. R., Hillel, D., and Benyamini, Y. (1970). Post irrigation movement of soil water: I. Redistribution, *Water Resources Res.* 6 (3), 851-861; II. Simultaneous redistribution and evaporation, *Water Resources Res.* 6 (4), 1148-1153.

Gingrich, J. R., and Russell, M. B. (1957). A comparison of the effects of soil moisture, tension and osmotic stress on root growth. *Soil Sci.* **84**, 185–191.

Graham, W. G., and King, K. M. (1961). Fraction of net radiation utilized in evapotranspiration from a corn crop. *Soil Sci. Soc. Amer. Proc.* **25**, 158–160.

Green, R. E., Hanks, R. J., and Larson, W. E. (1962). Estimates of field infiltration by numerical solution of the moisture flow equation. *Soil Sci. Amer. Proc.* **26**, 530–535.

Green, W. H., and Ampt, G. A. (1911). Studies on soil physics: I. Flow of air and water through soils. *J. Agr. Sci.* **4**, 1–24.

Greenland, D. J. (1965). Interaction between clays and organic compounds in soils. Part I. Mechanisms of interaction between clays and defined organic compounds. *Soil Fert.* **28**, 415–425.

Greenland, D. J., Lindstrom, G. R., and Quirk, J. P. (1962). Organic materials which stabilize natural soil aggregates. *Soil Sci. Soc. Amer. Proc.* **26**, 336–371.

Grim, R. E. (1953). "Clay Mineralogy." McGraw-Hill, New York.

Grim, R. E. (1958). Organization of water on clay mineral surfaces and its implications for the properties of claywater systems. Water and its Conduction in Soils. Highway Res. Board. Spec. Rep. 40. Nat. Acad. Sci.-Nat. Res. Council Publ. 629, pp. 17–23.

Guggenheim, E. A. (1959). "Thermodynamics." North-Holland, Amsterdam.

Gurr, C. G. (1962). Use of gamma-rays in measuring water content and permeability in unsaturated columns of soil. *Soil Sci.* **94**, 224–229.

Gurr, C. G., Marshall, T. J., and Hutton, J. T. (1952). Movement of water in soil due to temperature gradients. *Soil Sci.* **74**, 333–345.

Hadas, A., and Hillel, D. (1968). An experimental study of evaporation from uniform columns in the presence of a water table. *Trans. Int. Cong. Soil Sci., Adelaide 9th.* I: 67–74.

Hadley, G. (1961). "Linear Algebra." Addison-Wesley, Reading, Massachusetts.

Hagin, J., and Bodman, G. B. (1954). Influence of the polyelectrolyte CRD—186 on aggregation and other physical properties of some California and Israeli soils and some clay minerals. *Soil Sci.* **78**, 367–378.

Haines, W. B. (1930). Studies in the physical properties of soils. V. The hysteresis effect in capillary properties and the modes of moisture distribution associated therewith. *J. Agr. Sci.* **20**, 97–116.

Haise, H. R., Jensen, L. R., and Alessi, J. (1955). The effect of synthetic soil conditioners on soil structure and production of sugar beets. *Soil Sci. Amer. Proc.* **19**, 17–19.

Halstead, M. H., and Covey, W. (1957). Some meteorological aspects of evapotranspiration. *Soil Sci. Soc. Amer. Proc.* **21**, 461–464.

Hanks, R. J. (1960). Soil crusting and seedling emergence. *Trans. Int. Soil Sci. Congr. 7th* **I**, 340–346.

Hanks, R. J., and Bowers, S. B. (1962). Numerical solution of the moisture flow equation for infiltration into layered soils. *Soil Sci. Soc. Amer. Proc.* **26**, 530–534.

Hanks, R. J., and Gardner, H. R. (1965). Influence of different diffusivity-water content relations on evaporation of water from soils. *Soil Sci. Soc. Am. Proc.* **29**, 495–498.

Hanks, R. J., Gardner, H. R., and Fairbourn, M. L. (1967). Evaporation of water from soils as influenced by drying with wind or radiation. *Soil Sci. Soc. Amer. Proc.* **31**, 593–598.

Harrold, L. L. (1966). Measuring evapotranspiration by lysimetry. *In* " Evapotranspiration and its Role in Water Resources Management." Amer. Soc. Agr. Eng., St. Joseph, Michigan.

Hawkins, J. C. (1962). The effects of cultivation on aeration, drainage and other soil factors important in plant growth. *J. Sci. Food Agr.* **13**, 386–391.

Hendrickson, A. H., and Veihmeyer, F. J. (1945). Permanent wilting percentage of soils obtained from field and laboratory trials. *Plant Physiol.* **20**, 517–539.

Hide, J. C. (1954). Observation on factors influencing the evaporation of soil moisture. *Soil Sci. Soc. Amer. Proc.* **18**, 234–239.

Hide, J. C. (1958). Soil moisture conservation in the Great Plains. *Advan. Agron.* **10**, 23–36.

Hillel, D. (1960). Crust formation in loessial soils. *Trans. Int. Soil Sci. Congr. Madison 7th* **I**, 330–339.

Hillel, D. (1964). Infiltration and rainfall runoff as affected by surface crusts. *Trans. Int. Soil Sci. Cong. Bucharest. 8th* **2**, 53–62.

Hillel, D. (1968). Soil water evaporation and means of minimizing it. Rep. submitted to U.S. Dept. Agr. The Hebrew Univ. of Jerusalem, Israel.

Hillel, D., and Mottes, J. (1966). Effect of plate impedance, wetting method and aging on soil moisture retention *Soil Sci.* **102**, 135–140.

Hillel, D., and Gardner, W. R. (1969). Steady infiltration into crust topped profiles. *Soil Sci.* **108**, 137–142.

Hillel, D., and Gardner, W. R. (1970a). Transient infiltration into crust-topped profiles. *Soil Sci.* **109**, 69–76.

Hillel, D., and Gardner, W. R. (1970b). Measurement of unsaturated conductivity and diffusivity by infiltration through an impeding layer. *Soil Sci.* **109**.

Hillel, D., Rawitz, E., Schwartz, A., Shavit, A., and Steinhardt, R. (1967). Runoff inducement in arid lands. Res. Rep. submitted to U.S. Dept. of Agr. The Hebrew Univ. of Jerusalem, Israel.

Hillel, D., Gairon, S., Falkenflug, V., and Rawitz, E. (1969). New design of a low-cost hydraulic lysimeter for field measurement of evapotranspiration. *Israel J. Agr. Res.* **19**, 57–63.

Hillel, D., and Hadas, A. (1970). Isothermal drying of structurally layered soil profiles. *Soil Sci.* (in press).

Holmes, J. W. (1950). Calibration and field use of the neutron scattering method of measuring soil water content. *Aust. J. Appl. Sci.* **7**, 45–58.

Holmes, J. W., and Jenkinson, A. F. (1959). Techniques for using the neutron moisture water. *J. Agr. Eng. Res.* **4**, 100–109.

Holmes, J. W., Greacen, E. L., and Gurr, C. G. (1960). The evaporation of water from bare soils with different tilths. *Trans. Int. Cong. Soil Sci. Madison, 7th* **1**, 188–194.

Holmes, J. W., Taylor, S. A., and Richards, S. J. (1967). Measurement of soil water. *In* "Irrigation of Agricultural Lands," pp. 275–306. *Amer. Soc. Agron.* Monograph 11.

Hooghoudt, S. B. (1937). Bijdregen tot de kennis van eenige natuurkundige grootheden van den grond, 6. *Versl. Landb. Ond.* **43**, 461–676.

Hornberger, G. M., Remson, I., and Fungaroli, A. A. (1969). Numeric studies of a composite soil moisture ground-water system. *Water Resources Res.* **5**, 797–802.

Horton, R. E. (1940). Approach toward a physical interpretation of infiltration capacity. *Soil Sci. Soc. Amer. Proc.* **5**, 339–417.

Hubbert, M. K. (1956). Darcy's law and the field equations of the flow of underground fluids. *Amer. Inst. Mining Met. Petrol. Eng. Trans.* **207**, 222–239.

Hutchinson, H. P., Dixon, I. S., and Denbigh, K. G. (1948). The thermo-osmosis of liquids through porous materials. *Discuss. Faraday Soc.* **3**, 86–94.

Isherwood, J. D. (1959). Water-table recession in tile-drained land. *J. Geophys. Res.* **64**, 795–804.

Israelsen, O. W., and West, F. L. (1922). Water holding capacity of irrigated soils. Utah State Univ. Agr. Exp. Sta. Bull. 183.

Israelsen, O. W., and Hansen, V. E. (1962). "Irrigation Principles and Practices." Wiley, New York.

Jackson, M. L., Tyler, S. A., Willis, A. L., Bourbeau, G. A., and Pennington, R. P. (1948). Weathering sequence of clay-size minerals in soils and sediments. I. Fundamental generalization. *J. Phys. Colloid Chem.* **52**, 1237–1260.

Jackson, R. D. (1964a). Water vapor diffusion in relatively dry soil: I. Theoretical considerations and sorption experiments. *Soil Sci. Soc. Amer. Proc.* **28**, 172–176.

Jackson, R. D. (1964b). Water vapor diffusion in relatively dry soil: II. Desorption experiment. *Soil Sci. Soc. Amer. Proc.* **28**, 464–466.

Jackson, R. D. (1964c). Water vapor diffusion in relatively dry soil: III. Steady state experiments. *Soil Sci. Soc. Amer. Proc.* **28**, 466–470.

Jackson, R. D., and Whisler, F. D. (1970). Approximate equations for vertical nonsteady-state drainage: I. Theoretical approach. *Soil Sci. Soc. Amer. Proc.* (in press).

Jacob, C. E. (1946). *Trans. Amer. Geophys. Union* **27**, 198.

Janert, H. (1934). The application of heat of wetting measurements to soil research problems. *J. Agr. Sci.* **24**, 136–145.

Jenny, H. (1932). Studies on the mechanism of ionic exchange in colloidal aluminum silicates. *J. Phys. Chem.* **36**, 2217–2258.

Jenny, H. (1935). The clay content of the soil as related to climatic factors, particularly temperature. *Soil Sci.* **40**, 111–128.

Jenny, H. (1938). "Properties of Colloids." Stanford Univ. Press, Palo Alto, California.

Jenny, H., and Reitemeier, R. F. (1935). Ionic exchange in relation to the stability of colloidal systems. *J. Phys. Chem.* **39**, 593–604.

Jensen, M. E., and Hanks, R. J. (1967). Nonsteady state drainage from porous media. *J. Irrig. Drain, Div. Amer. Soc. Civil Eng.* **93**, IR 3, 209–231.

Johnson, A. I. (1962). Methods of measuring soil moisture in the field. U.S. Dept. of Interior, Geological Survey Water Supply paper 1619-U.

Johnson, H. P., Frevert, R. K., and Evans, D. D. (1952). Simplified procedure for the measurement and computation of soil permeability below the water table. *Agr. Eng.* **33**, 283–289.

Katchalsky, A., and Curran, P. F. (1965). "Nonequilibrium Thermodynamics in Biophysics." Harvard Univ. Press, Cambridge, Massachussets.

Kavanu, J. L. (1964). "Water and Solute-Water Interactions." Holden-Day, San Fransisco, California.

Kemper, W. D. (1960). Water and ion movement in thin films as influenced by the electrostatic charge and diffuse layer of cations associated with clay mineral surfaces. *Soil Sci. Soc. Amer. Proc.* **24**, 10–16.

Kemper, W. D. (1965). Aggregate stability. *In* "Methods of Soil Analysis," pp. 511–519. Amer. Soc. Agron. Monograph 9.

Kemper, W. D., and Chepil, W. S. (1965). Size distribution of aggregates. *In* "Methods of Soil Analysis," pp. 499–510. Amer. Soc. Agron. Monograph 9.

King, K. M., Tanner, C. B., and Suomi, V. E. (1956). A floating lysimeter and its evaporation recorder. *Trans. Amer. Geophys. Union* **37**, 738–742.

Kirkham, D. (1957). The ponded water case. *In* "Drainage of Agricultural Lands," pp. 139–180. Amer. Soc. Agron. Monograph 7.

Kirkham, D. (1958). Seepage of steady rainfall through soil into drains *Trans. Am. Geophys. Union*, **32**, 892–908.

Kirkham, D., and Gaskell, R. E. (1957). The falling water table in tile and ditch drainage. *Soil Sci. Soc. Amer. Proc.* **15**, 37–42.

Klute, A. (1952). A numerical method for solving the flow equation for water in unsaturated materials. *Soil Sci.* **73**, 105–116.

Klute, A. (1965a). Laboratory measurement of hydraulic conductivity of saturated soil. *In* "Methods of Soil Analysis," pp. 210–221. Amer. Soc. Agron. Monograph 9.

Klute, A. (1965b). Laboratory measurement of hydraulic conductivity of unsaturated soil. *In* "Methods of Soil Analysis," pp. 253–261. Amer. Soc. Agron. Monograph 9.

Klute, A. (1965c). Water diffusivity. *In* "Methods of Soil Analysis," pp. 262–272. Amer. Soc. Agron. Monograph 9.

Klute, A. (1965d). Water capacity. *In* "Methods of Soil Analysis." pp. 273–278. Amer. Soc. Agron. Monograph 9.

Klute, A., Whisler, F. D., and Scott, E. J. (1965). Numerical solution of the flow equation for water in a horizontal finite soil column. *Soil Sci. Soc. Amer. Proc.* **29**, 353–358.

Kohnke, H., and Bertrand, A. R. (1959). "Soil Conservation." McGraw-Hill, New York.

Kostiakov, A. N. (1932). On the dynamics of the coefficient of water-percolation in soils and on the necessity for studying it from a dynamic point of view for purposes of amelioration. *Trans. Comm. Intern. Soil Sci. Soc., Moscow, 6th* Part A, 17–21.

Kozlowski, T. T., ed. (1968). "Water Deficits and Plant Growth," Vols. I and II. Academic Press, New York.

Kramer, P. J. (1956). Physical and physiological aspects of water absorption. *In* "Handbuch der Pflanzenphysiologie," Vol. III, Pflanze und Wasser, pp. 124–159. Springer-Verlag, Berlin.

Kramer, P. J., and Coile, T. S. (1940). An estimation of the volume of water made available by root extension. *Plant Physiol.* **15**, 743–749.

Kroth, E. M., and Page, J. B. (1947). Aggregate formation in soils with special reference to cementing substances. *Soil Sci. Soc. Amer. Proc.* **11**, 27–34.

Kruyt, H. R. (1949). "Colloid Science." Vols. 1 and 2. Elsevier, London.

Kunze, R. J., and Kirkham, D. (1962). Simplified accounting for membrane impedance in capillary conductivity determinations. *Soil Sci. Soc. Amer. Proc.* **26**, 421–426.

Langmuir, I. (1918). The adsorption of gases on plane surfaces of glass, mica and platinum. *J. Amer. Chem. Soc.* **40**, 1361–1402.

Lemon, E. R. (1956). The potentialities for decreasing soil moisture evaporation loss. *Soil Sci. Soc. Amer. Proc.* **20**, 120–125.

Lemon, E. R. (1960). Photosynthesis under field conditions. II. An aerodynamic method for determining the turbulent carbon dioxide exchange between the atmosphere and a corn field. *Agron. J.* **52**, 697–703.

Leonard, R. A., and Low, P. F. (1962). A self-adjusting nullpoint tensiometer. *Soil Sci. Soc. Amer. Proc.* **26**, 123–125.

Lippincot, E. R., Stromberg, R. R., Grant, W. H., and Cessac, G. L. (1969). Polywater. *Science* **164**, 1482–1487.

Low, P. F. (1951). Force fields and chemical equilibrium in heterogeneous systems with special reference to soils. *Soil Sci.* **71**, 409–418.

Low, P. F. (1961). Physical chemistry of clay-water interactions. *Advan. Agron.* **13**, 269–327.

Low, P. F. (1962). Influence of adsorbed water on exchangeable ion movement. *Clays, Clay Minerals* **9**, 219–228.

Low, P. F. (1968). Mineralogical data requirements in soil physical investigations. *In* "Mineralogy in Soil Science and Engineering," pp. 1–34. Soil Sci. Soc. Amer. Spec. Publ. No. 3.

Low, P. F. (1970). Hydrogen bonding and polywater in claywater systems. *Clays, Clay Minerals* (in press).

Low, P. F., and Deming, J. M. (1953). Movement and equilibrium of water in heterogeneous systems with special reference to soils. *Soil Sci.* **75**, 187–202.

Luthin, J. N., (ed.) (1957). "Drainage of Agricultural Lands." Amer. Soc. Agron. Monograph 7.

Luthin, J. N. (1966). "Drainage Engineering," Wiley, New York.

Lutz, J. F. (1937). The relation of free iron in the soil to aggregation. *Soil Sci. Soc. Amer. Proc.* **1**, 43–45.

Maasland, M., and Kirkham, D. (1955). Theory and measurement of anisotropic air permeability in soil. *Soil Sci. Soc. Amer. Proc.* **19**, 395–400.

Marshall, C. E. (1964). "The Physical Chemistry and Mineralogy of Soils." Wiley, New York.

Marshall, T. J. (1958). A relation between permeability and size distribution of pores. *J. Soil Sci.* **9**, 1–8.

Marshall, T. J. (1959). The diffusion of gases through porous media. *J. Soil Sci.* **10**, 79–82.

Martin, W. P., Taylor, G. S., Enjibous, J. C., and Burnett, E. (1952). Soil and crop responses from field applications of soil conditioners. *Soil Sci.* **73**, 455–471.

McIlroy, I. C., and Angus, D. E. (1963). The aspendale multiple weighed lysimeter installation. CSIRO Div. Meteorol. Phys. Tech. Paper No. 14. Melbourne, Australia.

McIntyre, D. S. (1956). The effect of free ferric oxide on the structure of some terra rossa and rendzina soils. *J. Soil Sci.* **7**, 302–306.

McIntyre, D. S. (1958). Soil splash and the formation of surface crusts by raindrop impact. *Soil Sci.* **85**, 261–266.

Millar, C. E., Turk, L. M., and Foth, H. D. (1965). "Fundamentals of Soil Science." Wiley, New York.

Miller, E. E., and Gardner, W. H. (1962). Water infiltration into stratified soil. *Soil Sci. Soc. Amer. Proc.* **26**, 115–118.

Miller, E. E., and Miller, R. D. (1955a). Theory of capillary flow: I. Practical implications. *Soil Sci. Soc. Amer. Proc.* **19**, 267–271.

Miller, E. E., and Miller, R. D. (1955b). Theory of capillary flow: II. Experimental information. *Soil Sci. Soc. Amer. Proc.* **19**, 271–275.

Miller, E. E., and Miller, R. D. (1956). Physical theory for capillary flow phenomena. *J. Appl. Phys.* **27**, 324–332.

Miller, E. E., and Elrick, D. E. (1958). Dynamic determination of capillary conductivity extended for non-negligible membrane impedance. *Soil Sci. Soc. Amer. Proc.* **22**, 483–486.

Miller, E. E., and Klute, A. (1967). Dynamics of soil water. Part I—Mechanical forces. *In* "Irrigation of Agricultural Lands," pp. 209–244. Amer. Soc. Agron. Monograph 11.

Miller, R. D. (1951). A technique for measuring tension in rapidly changing systems. *Soil Sci.* **72**, 291–301.

Miller, R. J., and Low, P. F. (1963). Threshold gradient for water flow in clay systems. *Soil Sci. Soc. Amer. Proc.* **27**, 605–609.

Millington, R. J., and Quirk, J. P. (1959). Permeability of porous media. *Nature* **183**, 387–388.

Molz, F. J., Ramson, I., Fungaroli, A. A., and Drake, R. L. (1968). Soil moisture availability for transpiration. *Water Resources Res.* **4**, 1161–1169.

Moore, R. E. (1939). Water conduction from shallow water tables. *Hilgardia* **12**, 383–426.

Morgan, J., and Warren, B. E. (1938). X-ray analysis of the structure of water. *J. Chem. Phys.* **6**, 666–673.

Mortland, M. M., and Kemper, W. D. (1965). Specific surface. *In* "Methods of Soil Analysis," pp. 532–544. Amer. Soc. Agron. Monograph 9.

Muskat, M. (1946). "The Flow of Homogeneous Fluids Through Porous Media." Edwards, Ann Arbor, Michigan.

Narten, A. H., and Levy, H. A. (1969). Observed diffraction pattern and proposed model of liquid water. *Science* **165**, 447–454.

Némethy, G., and Scheraga, H. A. (1962). Structure of water and hydraulic bonding in proteins. I. A model for the thermodynamic properties of liquid water. *J. Chem. Phys.* **36**, 3382–3400.

Nerpin, S., Pashkina, S., and Bondarenko, N. (1966). The evaporation from bare soil and the way of its reduction. Symp. Water in the Unsaturated Zone, Wageningen.

Nielsen, D. R., and Biggar, J. W. (1961). Miscible displacement in soils. I. Experimental information. *Soil Sci. Soc. Amer. Proc.* **25**, 1–5.

Nielsen, D. R., and Biggar, J. W. (1963). Miscible displacement: Mixing in glass beads. *Soil Sci. Soc. Amer. Proc.* **27**, 10–13.

Nielsen, D. R., Biggar, J. W., and Davidson, J. M. (1962). Experimental consideration of diffusion analysis in unsaturated flow problems. *Soil Sci. Soc. Amer. Proc.* **26**, 107–112.

Nixon, P. R., and Lawless, G. P. (1960). Translocation of moisture with time in unsaturated soil profiles. *J. Geophys. Res.* **65**, 655–661.

Ogata, G., and Richards, L. A. (1957). Water content changes following irrigation of bare field soil that is protected from evaporation. *Soil Sci. Soc. Amer. Proc.* **21**: 355–356.

Ogata, G., Richards, L. A., and Gardner, W. R. (1960). Transpiration of alfalfa determined from soil water content changes. *Soil Sci.* **89**, 179–182.

Olsen, H. W. (1965). Deviations from Darcy's law in saturated clays. *Soil Sci. Soc. Amer. Proc.* **29**, 135–140.

Olsen, S. R., and Kemper, W. D. (1968). Movement of nutrients to plant roots. Advan. Agron. **20**, 91–151.

Parr, J. F., and Bertrand, A. R. (1960). Water infiltration into soils. *Advan. Agron.* **12**, 311–363.

Pauling, L. (1960). "The Nature of the Chemical Bond." Cornell Univ. Press, Ithaca, New York.

Peaceman, D. W., and Rachford, H. H., Jr. (1955). The numerical solution of parabolic and elliptic differential equations. *J. Soc. Ind. Appl. Math.* **3**, 28–41.

Pearse, J. F., Oliver, T. R., and Newitt, D. M. (1949). The mechanism of the drying of solids: Part I. The forces giving rise to movement of water in granular beds during drying. *Trans. Inst. Chem. Eng. (London)* **27**, 1–8.

Peck, A. J. (1960). The water table as affected by atmospheric pressure. *J. Geophys. Res.* **65**, 2383–2388.

Peck, A. J. (1970). Redistribution of soil water after infiltration. Australian *J. Soil Res.* (in press).

Peck, A. J., and Rabbidge, R. M. (1969). Design and performance of an osmotic tensiometer for measuring capillary potential. *Soil Sci. Soc. Amer. Proc.* **33**, 196–202.

Pelton, W. L. (1961). The use of lysimetric methods to measure evapotranspiration. *Proc. Hydrol. Symp.* **2**, 106–134 (Queen's Printer, Ottawa, Canada. Cat. No. R32–361/2.)

Penman, H. L. (1940). Gas and vapour movements in the soil. I. The diffusion of vapours through porous solids. *J. Agr. Sci.* **30**, 437–462.

Penman, H. L. (1948). Natural evaporation from open water, bare soil and grass. *Proc. Roy. Soc. London* **A193**, 120–146.

Penman, H. L. (1949). The dependence of transpiration on weather and soil conditions. *J. Soil Sci.* **1**, 74–89.

Penman, H. L. (1956). Evaporation: an introductory survey. *Neth. J. Agr. Sci.* **4**, 9–29.

Philip, J. R. (1955a). Numerical solution of equations of the diffusion type with diffusivity concentration dependent. *Trans. Faraday Soc.* **51**, 885–892.

Philip, J. R. (1955b). The concept of diffusion applied to soil water. *Proc. Nat. Acad. Sci. (India)* **24A**, 93–104.

Philip, J. R. (1957a). Numerical solution of equations of the diffusion type with diffusivity concentration—dependent: 2. *Aust. J. Phys.* **10**, 29–42.

Philip, J. R. (1957b). Evaporation, moisture and heat fields in the soil. *J. Meteorol.* **14**, 354–366.

Philip, J. R. (1957c). The physical principles of soil water movement during the irrigation cycle. *Proc. Intern. Cong. Irrig. Drainage, 3rd,* **8**, 125–8, 154.

Philip, J. R. (1957d). The theory of infiltration: 1: The infiltration equation and its solution. *Soil Sci.* **83**, 345–357.

Philip, J. R. (1957e). The theory of infiltration: 2: The profile at infinity. *Soil Sci.* **83**, 435–448.

Philip, J. R. (1957f). The theory of infiltration: 3: Moisture profiles and relation to experiment. *Soil Sci.* **84**, 163–178.

Philip, J. R. (1957g). The theory of infiltration: 4: Sorptivity and algebraic infiltration equations. *Soil Sci.* **84**, 257–264.

Philip, J. R. (1957h). The theory of infiltration: 5: The influence of initial moisture content. *Soil Sci.* **84**, 329–339.

Philip, J. R. (1957i). The theory of infiltration: 6: Effect of water depth over soil. *Soil Sci.* **85**, 278–286.

Philip, J. R. (1960). Absolute thermodynamic functions in soil-water studies. *Soil Sci.* **89**, 111.

Philip, J. R. (1964). Similarity hypothesis for capillary hysteresis in porous materials. *J. Geophys. Res.* **69**, 1553–1562.

Philip, J. R. (1966). Plant water relations: Some physical aspects. *Ann. Rev. Plant Physiol.* **17**, 245–268.

Philip, J. R. (1969a). Hydrostatics and hydrodynamics in swelling soils. *Water Resources Res.* **5**, 1070–1077.

Philip, J. R. (1969b). Theory of infiltration. *Advan. Hydroscience* **5**, 216–296.

Philip, J. R. (1970). Flow in porous media. *Ann. Rev. Fluid Mech.* **2** (in press).

Philip, J. R., and deVries, D. A. (1957). Moisture movement in porous materials under temperature gradients. *Trans. Amer. Geophys. Union* **38**, 222–228.

Phillips, R. E., and Kirkham, D. (1962). Mechanical impedance and corn seedling root growth. *Soil Sci. Soc. Amer. Proc.* **26**, 319–322.

Pillsbury, A. F. (1950). *Soil Sci.* **70**, 299.

Porter, K. K., Kemper, W. D., Jackson, R. D., and Stewart, B. A. (1960). Chloride diffusion in soils as influenced by moisture content. *Soil Sci. Soc. Amer. Proc.* **24**, 460–463.

Poulovassilis, A. (1962). Hysteresis of pore water, an application of the concept of independent domains. *Soil Sci.* **93**, 405–412.

Poulovassilis, A. (1969). The effect of hysteresis of pore water on the hydraulic conductivity. *J. Soil Sci.* **20**, 52–56.

Prigogine, I. (1961). "Introduction to Thermodynamics of Irreversible Processes." Wiley, New York.

Pruitt, W. O., and Angus, D. E. (1960). Large weighing lysimeter for measuring evapotranspiration. *Trans. Amer. Soc. Agr. Eng.* **3**, 3–15, 18.

Quastel, J. H. (1954). Soil conditioners. *Res. Plant Physiol.* **5**, 75–92.

Quirk, J. P., and Schofield, R. K. (1955). The effect of electrolyte concentration on soil permeability. *J. Soil Sci.* **6**, 163–178.

Rawlins, S. L. (1966). Theory for thermocouple psychrometers used to measure water potential in soil and plant samples. *Agr. Meteorol.* **3**, 293–310.

Rawlins, S. L., and Dalton, F. N. (1967). Psychrometric measurement of soil water potentail without precise temperature control. *Soil Sci. Amer. Proc.* **31**, 297–301.

Reeve, R. C. (1957). Factors which affect permeability. *In* "Drainage of Agricultural Land," pp. 404–414. Amer. Soc. Agron. Monograph 7.

Reeve, R. C. (1965). Air-to-water permeability ratio. *In* "Methods of Soil Analysis," pp. 520–531. Amer. Soc. Agron. Monograph 9.

Reeve, R. C., Bower, C. A., Brooks, R. H., and Gschwend, F. B. (1954). A comparison of the effects of exchangeable sodium and potassium upon the physical condition of soils. *Soil Sci. Soc. Amer. Proc.* **18**, 130–132.

Reid, C. E. (1960). "Principles of Chemical Thermodynamics." Reinhold, New York.

Remson, I., Drake, R. L., McNeary, S. S., and Walls, E. M. (1965). Vertical drainage of an unsaturated soil. *Amer. Soc. Civil Eng. Proc. J. Hydraulics Div.* **91**, 55–74.

Remson, I., Fungaroli, A. A., and Hornberger, G. M. (1967). Numerical analysis of soil moisture systems. *Amer. Soc. Civil Eng. Proc. J. Irrig. Drain. Div.* **3**, 153–166.

Richards, L. A. (1931). Capillary conduction of liquids in porous mediums. *Physics* **1**, 318–333.

Richards, L. A., (ed.) (1954). Diagnosis and improvement of saline and alkali soils. U.S. Dept. Agr. Agr. Handbook 60.

Richards, L. A. (1955). Retention and transmission of water in soil. U.S. Dept. Agr. Yearbook Agr. 1955. pp. 144–151.

Richards, L. A. (1960). Advances in soil physics. *Trans. Intern. Congr. Soil Sci. Madison, 7th* I, 67–69.

Richards, L. A. (1965). Physical condition of water in soil. *In* "Methods of Soil Analysis" pp. 128–152. Amer. Soc. Agron. Monograph 9.

Richards, L. A., and Weaver, L. R. (1944). Fifteen atmosphere percentage as related to the permanent wilting percentage. *Soil Sci.* **56**, 331–339.

Richards, L. A., and Campbell, R. B. (1949). The freezing point of moisture in soil cores. *Soil Sci. Soc. Amer. Proc.* **13**, 70–74.

Richards, L. A., and Moore, D. C. (1952). Influence of capillary conductivity and depth of wetting on moisture retention in soil. *Trans. Amer. Geophys. Union* **33**, 4.

Richards, L. A., and Wadleigh, C. H. (1952). Soil water and plant growth. *In* "Soil Physical Conditions and Plant Growth," p. 13. Amer. Soc. Agron. Monograph 2.

Richards, L. A., and Ogata, G. (1958). A thermocouple for vapor pressure measurement in biological and soil systems at high humidity. *Science* **128**, 1084–1090.

Richards, L. A., Gardner, W. R., and Ogata, G. (1956). Physical processes determining water loss from soil. *Soil Sci. Soc. Amer. Proc.* **20**, 310–314.

Richards, S. J. (1965). Soil suction measurements with tensiometers. *In* " Methods of Soil Analysis," pp. 153–163. Amer. Soc. Agron. Monograph 9.

Richards, S. J., and Marsh, A. W. (1961). Irrigation based on soil suction measurements. *Soil Sci. Soc. Proc. Amer.* **25**, 65–69.

Richtmyer, R. D., and Morton, K. W. (1967). "Difference Methods for Initial-Value Problems," 2nd ed. Wiley (Interscience), New York.

Rijtema, P. E. (1959). Calculation of capillary conductivity from pressure plate outflow data with non-negligible membrane impedance. *Neth. J. Agr. Sci.* **7**, 209–215.

Robins, J. S., Pruitt, W. O., and Gardner, W. H. (1954). Unsaturated flow of water in field soils and its effect on soil moisture investigations. *Soil Sci. Soc. Amer. Proc.* **18**, 344–348.

Romkens, M. J. M., and Bruce, R. R. (1964). Nitrate diffusivity in relation to moisture content of nonadsorbing porous media. *Soil Sci.* **98**, 332–337.

Rose, C. W. (1966). "Agricultural Physics." Pergamon Press, Oxford.

Rose, C. W. (1968). Evaporation from bare soil under high radiation conditions. *Trans. Intern. Cong. Soil Sci., Adelaide, 9th,* I, 57–66.

Rose, C. W. (1968). Water transport in soil with a daily temperature wave: I. Theory and experiment. *Aust. J. Soil Res.* **6**.

Rose, C. W. (1968). Water transport in soil with a daily temperature wave: II. Analysis. *Aust. J. Soil Res.* **6**.

Rose, C. W., and Stern, W. R. (1967a). Determination of withdrawal of water from soil by crop roots as a function of depth and time. *Aust. J. Soil Res.* **5**, 11–19.

Rose, C. W., and Stern, W. R. (1967b). The drainage component of the water balance equation. *Aust. J. Soil. Res* **3**, 95–100.

Rose, C. W., Stern, W. R., and Drummond, J. E. (1965). Determination of hydraulic conductivity as a function of depth and water content for soil *in situ. Aust. J. Soil Res.* **3**, 1–9.

Rose, C. W., Byrne, G. F., and Begg, J. E. (1966). An accurate hydraulic lysimeter with remote weight recording. CSIRO Div. Land Res. and Reg. Survey, Tech. Paper 27. Canberra, Australia.

Rose, D. A. (1966). Water transfer in soils by evaporation and infiltration. Symp. Water in the Unsaturated zone. Wageningen.

Rubin, J. (1966). Theory of rainfall uptake by soils initially drier than their field capacity and its applications. *Water Resources Res.* **2**, 739–749.

Rubin, J. (1967). Numerical method for analyzing hysteresis-affected, post-infiltration redistribution of soil moisture. *Soil Sci. Soc. Amer. Proc.* **31**, 13–20.

Rubin, J. (1968). Theoretical analysis of two-dimensional transient flow of water in un-saturated and partly unsaturated soils. *Soil Sci. Soc. Amer. Proc.* **32**, 607–615.

Rubin, J., and Steinhardt, R. (1963). Soil water relations during rain infiltration: I. Theory. *Soil Sci. Soc. Amer. Proc.* **27**, 246–251.

Rubin, J., and Steinhardt, R. (1964). Soil water relations during rain infiltration: III. Water uptake at incipient ponding. *Soil Sci. Soc. Amer. Proc.* **28**, 614–619.

Rubin, J., Steinhardt, R., and Reiniger, P. (1964). Soil water relations during rain infiltration: II. Moisture content profiles during rains of low intensities. *Soil Sci. Soc. Amer. Proc.* **28**, 1–5.

Russell, E. W. (1938). Soil structure. Imperial Bur. Soil Sci. Tech. Commun. 317.

Russell, M. B. (1952). Soil aeration and plant growth. *In* "Soil Physical Conditions and Plant Growth," pp. 253–301. Amer. Soc. Agron. Monograph 2.

Russell, M. B., and Feng, C. L. (1947). Characterization of the stability of soil aggregates. *Soil Sci.* **63**, 299–304.

Scheidegger, A. E. (1957). "The Physics of Flow through Porous Media." Macmillan, New York.

Schofield, R. K. (1935). The pF of the water in soil. *Trans. Intern. Cong. Soil Sci. 3rd* **2**, 37–48.

Sellers, W. D. (1965). "Physical Climatology." Univ. of Chicago Press, Chicago, Illinois.

Shaw, B. T., (ed.) (1952). "Soil Physical Conditions and Plant Growth." Academic Press, New York.

Slater, P. J., and Williams, J. B. (1965). The influence of texture on the moisture characteristics of soils. I. A critical comparison of techniques for determining the available water capacity and moisture characteristic curve of a soil. *J. Soil Sci.* **16**, 1–12.

Slatyer, R. O. (1967). "Plant-Water Relationships." Academic Press, London.

Slichter, C. S. (1899). U.S. Geol. Surv. Ann. Rep. 19—II: 295–384.

Smiles, D. E., and Rosenthal, M. J. (1968). The movement of water in swelling materials. *Aust. J. Soil Res.* **6**, 237–248.

Smith, G. D. (1965). "Numerical Solution of Partial Differential Equations." Oxford Univ. Press, London and New York.

Smythe, W. R. (1950). "Static and Dynamic Electricity." McGraw-Hill, New York.

Soil (1957). Yearbook of Agriculture, U.S. Dept. of Agriculture.

Soil Survey Manual (1951). U.S. Dept. of Agr. Handbook. No. 18.

Sor, K., and Kemper, W. D. (1959). Estimation of surface area of soils and clays from the amount of adsorption and retention of ethylene glycol. *Soil Sci. Soc. Amer. Proc.* **23**, 105–110.

Southwell, R. V. (1946). "Relaxation Methods in Engineering Science." Oxford Univ. Press, London and New York.

Staple, W. J. (1969). Comparison of computed and measured moisture redistribution following infiltration. *Soil Sci. Soc. Amer. Proc.* (in press).

Stokes, G. G. (1851). On the effect of the internal friction of fluids on the motion of pendulums. *Trans. Cambridge Phil. Soc.* **9**, 8–106.

Stolzy, L. H., Weeks, L. V., Szuszkiewicz, T. E., and Cahoon, G. A. (1959). Use of neutron equipment for estimating soil suction. *Soil Sci.* **88**, 313–316.

Stolzy, L. H., and Letey, J. (1964). Characterizing soil oxygen conditions with a platinum microelectrode. *Advan. Agron.* **16**, 249–279.

Stolzy, L. H., van Gundy, S. D., Laubanauskas, C. K., and Szuszkiewicz, T. E. (1963). Response of *Tylenchulus semipenetran* infected citrus seedlings to soil aeration and temperature. *Soil Sci.* **96**, 292–298.

Stotzky, G. (1965). Microbial respiration. *In* "Methods of Soil Analysis," pp. 1550–1569. Amer. Soc. Agron. Monograph 9.

Swaby, R. J. (1950). The influence of humus on soil aggregation. *J. Soil Sci.* **1**, 182–194.

Swartzendruber, D. (1962). Non Darcy behavior in liquid saturated porous media. *J. Geophys. Res.* **67**, 5205–5213.

Tackett, J. L., and Pearson, R. W. (1965). Some characteristics of soil crusts formed by simulated rainfall. *Soil Sci.* **99**, 407–413.

Takagi, S. (1960). Analysis of the vertical downward flow of water through a two-layered soil. *Soil Sci.* **90**, 98–103.

Talsma, T. (1960). Comparison of field methods of measuring hydraulic conductivity. *Trans. Cong. Irrigation Drainage* **4**, C145–C156.

Talsma, T. (1963). The control of saline ground water. *Med. Landb. Wageningen* **63**(10), 1–68.

Tanner, C. B. (1957). Factors affecting evaporation from plants and soils. *J. Soil Water Conserv.* **12**, 221–227.

Tanner, C. B. (1960). Energy balance approach to evapotranspiration from crops. *Soil Sci. Soc. Amer. Proc.* **24**, 1–9.

Tanner, C. B. (1968). Evaporation of water from plants and soil. *In* "Water Deficits and Plant Growth." Academic Press, New York.

Tanner, C. B., and Elrick, D. E. (1958). Volumetric porous (pressure) plate apparatus for moisture hysteresis measurements. *Soil Sci. Soc. Amer. Proc.* **22**, 575–576.

Tanner, C. B., and Pelton, W. L. (1960). Potential evapotranspiration estimates by the approximate energy balance method of Penman. *J. Geophys. Res.* **65**, 3391–3413.

Tanner, C. B., and Lemon, E. R. (1962). Radiant energy utilized in evaporation. *Agron. J.* **54**, 207–212.

Taylor, S. A., and Cary, J. W. (1960). Analysis of the simultaneous flow of water and heat or electricity with the thermodynamics of irreversible processes. *Trans. Intern. Cong. Soil Sci., Madison, 7th* I, 80–90.

Taylor, S. A., and Slatyer, R. O. (1960). Water-soil-plant relations terminology. *Trans. Intern. Congr. Soil Sci., Madison, 7th* I, 394–403.

Taylor, G. S., and Luthin, J. N. (1963). The use of electronic computers to solve subsurface drainage problems. *Hilgardia* **34**, 543–558.

Taylor, G. S., and Luthin, J. N. (1969). Computer methods for transient analysis of water-table aquifers. *Water Resources Res.* **5**, 144–152.

Terzaghi, K. (1951). "Theoretical Soil Mechanisms." Chapman and Hall, London.

Thorthwaite, C. W. (1948). An approach toward a rational classification of climate. *Geograph. Rev.* **38**, 55–94.

Tickell, F. A., and Hyatt, W. N. (1938). *Bull. Amer. Ass. Petrol. Geol.* **22**, 1272.

Topp, G. C. (1969). Soil water hysteresis measured in a sandy loam and compared with the hysteresis domain model. *Soil Sci. Soc. Amer. Proc.* **33**, 645–651.

Topp, G. C., and Miller, E. E. (1966). Hysteresis moisture characteristics and hydraulic conductivities for glass-bead media. *Soil Sci. Soc. Amer. Proc.* **30**, 156–162.

van Bavel, C. H. M. (1965). Composition of soil atmosphere. *In* "Methods of Soil Analysis," pp. 315–318. Amer. Soc. Agron. Monograph 9.

van Bavel, C. H. M., and Myers, L. E. (1962). An automatic weighing lysimeter. *Agr. Eng.* **43**, 580–583.

van Bavel, C. H. M., Underwood, N., and Swanson, R. W. (1956). Soil moisture measurement by neutron moderation. *Soil Sci.* **82**, 29–41.

van Bavel, C. H. M., Stirk, G. B., and Brust, K. J. (1968a). Hydraulic properties of a clay loam soil and the field measurement of water uptake by roots: I. Interpretation of water content and pressure profiles. *Soil Sci. Soc. Amer. Proc.* **32**, 310–317.

van Bavel, C. H. M., Brust, K. J., and Stirk, G. B. (1968b). Hydraulic properties of a clay loam soil and the field measurement of water uptake by roots: II. The water balance of the root zone. *Soil Sci. Soc. Amer. Proc.* **23**, 317–321.

van der Molen, W. H. (1956). Desalinization of saline soils as a column process. *Soil Sci.* **81**, 19–27.

van Schilfgaarde, J. (1957). Approximate solutions to drainage flow problems. *In* "Drainage of Agricultural Lands," pp. 79–112. Amer. Soc. Agron. Monograph 7.

Varga, R. S. (1962). "Matrix Iterative Analysis." Prentice Hall, Englewood Cliffs, New Jersey.

Veihmeyer, F. J., and Hendrickson, A. H. (1927). Soil moisture conditions in relation to plant growth. *Plant Physiol.* **2**, 71–78.

Veihmeyer, F. J., and Hendrickson, A. H. (1931). The moisture equivalent as a measure of the field capacity of soils. *Soil Sci.* **32**, 181–193.

Veihmeyer, F. J., and Hendrickson, A. H. (1949). Methods of measuring field capacity and wilting percentages of soils. *Soil Sci.* **68**, 75–94.

Veihmeyer, F. J., and Hendrickson, A. H. (1950). Soil moisture in relation to plant growth. *Ann. Rev. Plant Physiol.* **1**, 285–304.

Veihmeyer, F. J., and Hendrickson, A. H. (1955). Does transpiration decrease as the soil moisture decreases? *Trans. Amer. Geophys. Union* **36**, 425–448.

Viets, F. G., Jr. (1962). Fertilizers and the efficient use of water. *Advan. Agron.* **14**, 228-261.

Vilain, M. (1963). L'aeration du sol. Mise au point bibliographique. *Ann. Agron.* **14**, 967–998.

Visser, W. C. (1959). Crop growth and availability of moisture. Inst. of Land and Water Management, Wageningen, Netherlands, Tech. Bull. No. 6.

Visser, W. C. (1966). Progress in the knowledge about the effect of soil moisture content on plant production. Inst. Land Water Management, Wageningen, Netherlands, Tech. Bull. 45.

Vomocil, J. A., and Flocker, W. J. (1961). Effect of soil compaction on storage and movement of soil air and water. *Trans. Amer. Soc. Agr. Engr.* **4**, 242–246.

United States Department of Agriculture (1955). Water. *In* "Yearbook of Agriculture," U.S. Government Printing Office, Washington, D.C.

Watson, K. K. (1966). An instantaneous profile method for determining the hydraulic conductivity of unsaturated porous materials. *Water Resources Res.* **2**, 709–715.

Weinberg, A. M., and Wigner, E. P. (1958). "The Physical Theory of Neutron Chain Reactors." Univ. of Chicago Press, Chicago, Illinois.

Whisler, F. D., and Klute, A. (1965). The numerical analysis of infiltration, considering hysteresis, into a vertical soil column at equilibrium under gravity. *Soil Sci. Soc. Amer. Proc.* **29**, 489–494.

Wiegand, C. L., and Taylor, S. A. (1961). Evaporative drying of porous media. Agr. Exp. Sta. Utah State Univ., Logan. Spec. Rep. 15.

Willis, W. O. (1960). Evaporation from layered soils in the presence of a water table. *Soil Sci. Soc. Amer. Proc.* **24**, 239–242.

Wind, G. P. (1955). Flow of water through plant roots. *Neth. J. Agr. Sci.* **3**, 259–264.

Winger, R. J. (1960). In-place permeability tests and their use in subsurface drainage. Off. of Drainage and Ground Water Eng., Bur. of Reclamation, Denver, Colorado.

Wolf, J. M. (1968). The role of root growth in supplying moisture to plants. Unpublished PhD Thesis. Univ. of Rochester.

Yoder, R. E. (1936). A direct method of aggregate analysis and a study of the physical nature of erosion losses. *J. Amer. Soc. Agron.* **28**, 337–351.

Youngs, E. G. (1958a). Redistribution of moisture in porous materials after infiltration. *Soil Sci.* **86**, 117–125.

Youngs, E. G. (1958b). Redistribution of moisture in porous materials after infiltration. *Soil Sci.* **86**, 202–207.

Youngs, E. G. (1960a). The hysteresis effect in soil moisture studies. *Trans. Intern. Soil Sci. Cong. Madison. 7th* **1**, 107–113.

Youngs, E. G. (1960b). The drainage of liquids from porous materials. *J. Geophys. Res.* **65**, 4025–4030.

Youngs, E. G. (1964). An infiltration method of measuring the hydraulic conductivity of unsaturated porous materials. *Soil Sci.* **97**, 307–311.

Author Index

Numbers in italics refer to the pages on which the complete references are listed.

Subject Index

A

Adhesion, 80
Adsorbed water, 44, 70, 121
Adsorption, 21–24, 44, 57, 58, 77, 123
Advection of energy, 235–236
Aeration, soil, 125–126, 167
Aggregation, aggregates (clods), 21, 25,
    134, 140, 144, 152, 199
Air entrapment, 67, 68, 93
Air-entry value (of suction), 61, 107,
    134, 145
Air-filled porosity, 126
Albedo (reflectivity), 231, 232
Aluminosilicate minerals, 18–20
Anaerobic conditions, 167
Analog models, 177, 210, 244
Anistropy, *see* Isotropy and anisotropy
Artesian pressure, 178
Availability of soil water to plants, 63
    classical concepts, 202–206
    newer concepts, 206–207

B

Balance of energy in the field, 232–233
Balance of water in the field, 225–230
B.E.T. equation, adsorption, 24
Bingham liquid, 95
Boltzmann solution (transformation),
    114, 136, 255–257
Boundary layer (laminar & turbulent,
    208, 233
Bowen ratio, 234, 235, 237
Breakthrough curves, 123
Bulk density, 10–11

C

Capillarity, 37, 58, 65
Capillary conductivity, 109
Capillary fringe, 169
Capillary model, 185
Capillary potential, 50, 57
Capillary rise in soils, 168, 185, 186, 188,
    189, 191, 226
Capillary water, 50
Cation exchange, 19–20, 44
Characteristic curve of soil moisture,
    61–66
    measurement of, 76–77
Chemical potential, 54
Clausius-Clapeyron equation, 34
Clay domains, 25
Clay minerals, 20
Clay, properties and behavior, 18–21
    aluminosilicates, 18
    clay minerals, 18
    deflocculation, 17
Clay skins, 26
Clothesline effect, 235
Colloids, 14, 18–20
Compaction, 146, 168
Composite column, flow in, 89–90
Conductance, hydraulic, 220, 221
Contact angle (liquid on solid), 43, 61,
    66
Continuity equation, 100, 122, 137, 243
Coulomb forces, 44
Cross-coefficients, Onsager, 119
Crust, surface, 21, 26, 116, 133, 134, 140
    effect on infiltration, 144–149, 150
Cracked soil, 140, 185, 199